DESIGN *matters*

Overleaf
Heritage in cobblestones, contentment and confidence: The Leaches at Colne. A view dear to the hearts of Lancashire Conservationists and a widely held attitude of mind, unchanged since the Industrial Revolution, to the environment. (*Sunday Times*)

DESIGN *matters*

BERNAT KLEIN

London
SECKER & WARBURG

First published in England 1976 by
Martin Secker & Warburg Limited
14 Carlisle Street, London, W1V 6NN

SBN 436 23470 X

Designed by Ruari McLean Associates Limited,
Dollar, Scotland FK14 7PT

Filmset and printed in Great Britain by BAS Printers Limited
Wallop, Hampshire
and bound by W. & J. Mackay Limited
Chatham, Kent

Also by Bernat Klein
Eye for Colour

CONTENTS

I Why this book was written 1

II Definitions of design 3
The relevance of design to the life of the individual and to the economy of the nation
The Industrial Revolution and its effects on design
Some factors behind past successes of our exports
How it all changed

III Some comparisons between British and European attitudes to things seen 19
'Them and Us' on the shop-floor in British industry
Some other social and industrial attitudes which affect design and quality
Exports and their dependence on the quality of demand on the home market
What others, abroad, think of us

IIIA Scandinavia as I see it: 22 photographs 36

IV The 'We know what they want and they can jolly well have it' attitude of industry 52

IVA Britain as I see it: 32 photographs 60

V The 'Marks & Spencerisation' of Britain 77
Most people choose naturally well – if they have a choice

VI The difference between choosing and designing 83
On freedom of choice

VII Which comes first, demand or supply? 92
The visible and the hidden standards of demand
How education affects the standard of demand
The need for changes in our system of education

VIII 'Good Education' and visual literacy 97
An overall view of art and crafts as a subject in our schools

IX Creativity and sensitivity are different talents 105
A work sequence
Creativity
Pleasure seeking

X Campbell's work and theory of man as a pleasure-seeking animal 112
What children can and cannot be taught
The importance of a user-orientated teaching approach

XI Our primary schools today 120
The need for classes in environmental appreciation
The need for a new kind of teacher

XII Our secondary schools today 127
The interviews with teachers and some impressions

XIII Our art schools and schools of art and crafts 134
Their original purpose and evolution
Some of their current preoccupations
The 'generalist versus specialist' controversy
Conclusions and suggestions

XIIIA The output of our top schools of art: 10 photographs 144

XIV The Design Council 151
Its handicapping virtues
The Design Council at its own valuation (1972–73 Annual Report)
An analysis of what the Council does and stands for
Preaching to the converted and the making of instant converts
Conferences and other talking shops
Services to industry, awards, visitors
Policy on engineering products.
The Council's rôle in our educational system
Design magazine
Some of the Council's other virtues
Suggestions

XIVA Examples of Design Council award-winning products: 30 photographs over the past ten years 170

XIVB Some examples of good British Design 184

XV Conclusions 198

Appendices

A Statistics on schools, students and teachers 201

B Notes from *After Coldstream* (statistics and views) 205

C Secondary School Education. Summaries from reports and some views and conclusions 210

D Specialists versus generalists at our schools of art and crafts 223

E A summary of Stuart Macdonald's article on Articidal tendencies in *After Coldstream* 234

F The Design Council: its foundation and current performance 238

REFERENCES

The author and his publisher gratefully acknowledge permission for the quotations and references:

Books quoted

Industry And Empire. E. J. Hobsbawm (Weidenfeld & Nicolson 1968)
Chronicles of Wasted Time. M. Muggeridge (Collins 1972)
Religion and the Rise of Capitalism. R. H. Tawney (Murray 1926)
The Greening of America. C. A. Reich (Allen Lane The Penguin Press 1971)
The Pleasure Areas. H. J. Campbell (Eyre Methuen 1973)
Turn Again Westminster. Woodrow Wyatt (Deutsch 1973)
Readings in Art and Design Education: 1. After Hornsey; 2. After Coldstream. David Warren Piper (ed.) (Davis-Poynter 1973)

Books mentioned

The Fate of the Edsel and Other Business Adventures. John Brooks (Harper & Row, New York 1963)
Obedience to Authority. Stanley Milgram (Tavistock 1974)
The Collapse of British Power. Correlli Barnett (Eyre Methuen 1973)

Articles quoted

Article in *The Observer* by Michael Frayn, April 1974
Articles in *The Times*: Profile of Two Managing Directors of Marks & Spencer, ('Europa') 8 January 1974; by Ronald Butt, 7 February 1974; *The Times* Diary, 15 August 1975
Article in *Encounter* of Lecture to the British Academy 1973 – 'The Lion and the Ostrich', Arthur Koestler (by permission of A. D. Peters & Co Ltd)
Articles in *Encounter* by Theo Sommer (November 1973) and Max Beloff (November 1973)
Articles in *The Listener*: Kenneth Harris, 25 May 1972; Christopher Serpell, 19 July 1973
Article in *The Economist* by Norman Macrae, April 1973
Article in *The Economist*, 26 May 1975

Other sources

Extracts from HMSO Publications: *Art and Crafts in the Primary School* (1968); *Art Education* (1960); *Curriculum Paper No. 9 – Art in Secondary Schools* (1971)
Extracts from the Design Council's Annual Report for 1972/73
Extract from letter by Dr. P. W. Smith, President of the BTEA and Chairman of the Shirley Institute, in *The Textile Institute and Industry* (June 1973)

ACKNOWLEDGMENTS

I should like to thank Michael Laird, the Edinburgh architect, Ruari McLean, the book designer and writer, my colleague George Norwich, Gillian Colledge, and my son Jonathan, for their help and encouragement in the preparation of this book.

I also thank the teachers who were interviewed in the course of the preparation of this book and who helped me to form a clearer opinion of the problems we face in regard to design.

The points I put forward depend for their validity on my day-to-day work and on my experiences in many factories over a long number of years. I am therefore very grateful to the managers and other workers in industry, in Britain and abroad, who have shown great patience and given me their unstinted co-operation since I first started work as a designer.

My special thanks go to Howell Leadbeater, Controller of Supplies at the Department of the Environment, and to the executives in that Department. Throughout our work together they proved, with imagination, courage and perseverance that compromise in design is not an indispensable ingredient of success in practical, commercial terms; for that is, in a sense, what this book is about.

SOURCES OF ILLUSTRATIONS

The author and his publisher gratefully acknowledge permission to reproduce the illustrations in this book:

Barnaby's Picture Library *pp.* 16, 17, 62, 63, 64, 65, 66, 185
BBC Publications *p.* 67
Brecht-Einzig Ltd *pp.* 191, 196
British Railways Board *p.* 188
Camera Press *p.* 61
Central Office of Information *p.* 190
Central School of Art & Design *pp.* 145, 146
Crossley Carpets Ltd *p.* 171
Daily Mirror Newspapers Ltd *p.* 17
Daily Telegraph pp. 16, 21
Design Council, *Design* magazine *pp.* 58–9, 147, 148, 149, 150, 172, 173, 174, 175, 176, 177, 178, 179, 180, 181, 182, 183
Grey Lacey Photography *p.* 187
Heals Fabrics Ltd *pp.* 177, 182, 183
Jock Kinneir *p.* 197
Owen Lawrence Ltd *p.* 195
Edward Leigh *p.* 194
John Murray Ltd *p.* 20
National Film Archive *p.* 61
Philipson Studios *p.* 189
Punch p. 32
Royal College of Art *pp.* 145, 146
SIAD, *Designer pp.* 149, 176, 183
Sunday Telegraph p. 90
Times Newspapers Limited *pp.* i, 11, 109
Colin Westwood *p.* 194

I

Among the subjects which have fascinated me since my late teens there are two in which my interest seems, with the passing years, to have increased rather than diminished: in what things look like and in what the British are really like.

I have written this book, which deals with some aspects of these two subjects, because its writing gave me pleasure. I like to speculate about design and I enjoy trying to understand the teasingly elusive and intangible British attitudes to living in general, to appearances and to the way things look in particular.

Yet, these are not an onlooker's notes, detached and dispassionate. I am totally involved in the daily life and in the problems of this country and do not entirely disclaim the hope, perhaps even the ambition, of influencing the thinking of others through this book; of helping to change things. As a designer I consciously strive, through my work, to make people see things my way and, since I am writing here mainly about design, I do not pretend that I am indifferent to the readers' reactions and shall not try to hide my lack of bias.

Objectivity is a state of grace which I do not often attain. It is, however, one of the demands which my work, as a design consultant to companies in many countries, makes on me. When a new commission is being discussed I am usually asked to make an appraisal, as objective as possible, of the design aspects of the product in which the prospective client company is interested and of the relation of the design factors to the aims of the company. Some of the objectivity which has become an important part of my work as a designer has, I hope, seeped through into a basically subjective nature and will make what I have to say less lopsided.

My initial reports to clients are usually headed 'As I See It', a title which acknowledges the lack of complete impartiality and at the same time implies the possibility of other interpretations of the facts and admits the likelihood of other solutions to problems which all functioning enterprises face from time to time.

In a sense this is my 'As I See It' report on a Britain which has to earn its living largely by selling its artefacts.

II

My years in industry, on the shop floor and in the boardroom, have taught me that there are few words which are more ambiguous and more often misunderstood than 'design'. It means anything anybody wants it to mean and since few people who often use it explain what they mean by it, there is much confusion and little happiness at the end of many exercises involving 'design'.

As it is the primary and legitimate function of industry to make the highest profits in the shortest time involving the least risk and a minimum of capital, most manufacturers think 'copy' when they say 'design'. To me, however, to design means to enjoy the exploration of new possibilities. It means to take pleasure in finding new solutions to old problems; or to have fun juggling with a number of old solutions until they suddenly click and coalesce into one, beautiful, new solution.

Others have a mock-humble attitude to design. They say they know little about it and prefer to leave the choice to their customers. They pronounce themselves to be the 'servants' of their public and expect to be told what to produce. To me, however, to design means to initiate. It is not saying to a customer: 'We can and are prepared to do anything. You tell us what you want and we'll make it for you.' To design means to take a brief when necessary and at the same time to have a definite point of view of one's own in regard to the project and to put it to the client or customer in convincing terms.

Yet others adopt a superior but hidden tone vis-à-vis their public and are convinced that they themselves are so far ahead that the public can't follow them anyhow; so why try? Give the public what it wants instead. I am fairly certain, though, that designing industrially is not saying: 'We have some wonderful ideas but we are not going to do anything about them; the public isn't ready, wouldn't understand and appreciate what we have to offer.' To me, to design means not to underestimate the public. It means producing things even if they have never been – or perhaps just because they have never been – made and sold before.

Then, of course, most people dislike and are even a little afraid of change; but design means change. It is not looking up the sales ledger, seeing what sold last year and making something similar for next year. It means trying out something new.

There are those to whom design means finding a good, selling product, already well established in the market and then reproducing it with certain modifications so that it can be sold more cheaply. However I would not call this kind of work design, because design usually means paying the same for something better or paying more for something much better. It is not saying: 'Our competitors are producing something for so much; we could make it for 10, 20, 30% less.' This may be good raw material buying, good production planning, good business or it may be good throat-cutting. Design is not primarily concerned with these matters. Designing means knowing but forgetting what the competition is doing and doing something original where the price is not the main selling factor. It means backing one's judgment.

Designing could be described as the antithesis of tradition. This is not to condemn tradition totally in the sense of experiences lived through, classified, stored and made available to be drawn upon when necessary. After all this is what growing up is; maturity and an adult approach are as important in design as in any other sphere of human activity. But to accept another person's experiences, sieved through another personality into another's traditions is quite a different thing. To design means, therefore, to be aware of one's own experiences (and this includes legitimate effort to go through and to some extent make other people's experiences into one's own) but it means even more to work out each problem anew without taking for granted the validity of anything which already exists. In the sense that there are very few laws in aesthetics and because the main ingredients of artistic activity are freshness of approach, the free play of feelings, application and self-control, to rely mainly on tradition in design is unlikely to be productive.

Design is also a little aggressive. It should actively question accepted thought and dogma. It should reason instead of believe (even if some of the reasoning is done in the gut rather than in the head).

It follows that design is also the antithesis of the common

cliché, the hand-me-down, elastic wisdom of the conformist, the mentally timid or the lazy. For the cliché is the envelope for pigeonholing the strange, alarming idea or situation. The curious and the open-minded will relish the challenge of surprising new views and will enjoy pitching their mental resources against the unexpected. In the end they will be ready to exchange a view held for a better one if it comes along. The timid, by reflex reaction, reach for the cliché; it seems to help because it either pretends to identify the new with something from the past and thus reassures or, better still, it completely disposes of the challenge by implicit condemnation, usually on specious moral or ethical grounds.

Design is incompatible with another kind of cliché, almost never recognised as such, for this attitude, on the outside, seems to be positive, constructive and receptive to new ideas. It is the state of mind with which many trained in scientific and technological disciplines, regard their work and the world around them; facts and figures, charts and graphs are things which they understand and for which they have an insatiable appetite. Any amount of time spent on them, proving no matter what, is to them time well spent. They feel that functioning on a fare of graphs is safer than taking inspired leaps into the dark; that the meticulous assembly of tremulous trends contributes more to the solution of problems than speculation and inspiration. When confronted with an inexplicable situation they reach for their graphs. To them the clamping on of formulae to problems is original, constructive thinking.

The milestones of human progress, however, stand firmly on a foundation of imagination: on the conception of the inconceivable by scientists, philosophers and artists. It is on this foundation that others, with less original ideas, can build their systematic, perhaps mechanical, constructions. Certain clichés of scientific methodology must not be allowed to obstruct the true importance of the rôle of the imagination. Facts are sacred. Imagination can change them by creating new ones. Design, therefore, means the exercise and the application of imagination; a word that has lost much of its meaning but which, when freshly used, implies a flight of fantasy, the materialisation of intuition, the creation of something out of nothing.

Designers, not having the certainties of science and technology to support them in their doubts, need to believe in the importance of imagination.

Designing, therefore, implies an open, enquiring, searching mind combined with an adventurous attitude. The designer should be a visionary and a dreamer; the practical designer, the fool who dares; who attempts to translate vision into reality.

The thesis of design is the physical world as we know it – people and their environment as they are and as they could, with advantages all round, be. It is the attempt at the reconciliation of what we want with what we have; but more, it means making the best of what we have.

Design does not accept facts just because they are there. It is often concerned with the peeling away of the sticky layers of pretentiousness into which inert tradition tends to wrap succeeding generations. For among other things and despite appearances to the contrary, each generation inherits from the preceding one the sum total of its prejudices and pretentions. These gather and solidify around small and large things in life and obscure the central issues. In so far as it is the function of design to get to the heart of things, it has to be destructive, almost brutal, in its insistence on doing away with pretensions, irrelevancies and inessentials. Design has to reduce and simplify, to make things plainer rather than more complicated; the more successfully it does this the easier it will be to follow the work of the designer and the more people will be able to enjoy it. Design, whether from a moral or from a practical point of view, should be for all, not just for an élite.

Design tries to get to the truth by the shortest route. It recognises that there are many routes to it and when it gets there it does not say: 'This is the ultimate.' It merely suggests a closer approximation than existed before to the best solution of the problem in hand. The implication is clear that this too will one day be replaced by something better.

Good design respects its raw materials and recognises that it has to fulfil a practical as well as an aesthetic function; it reconciles raw material and form and function; function and colour and texture; and all of these with the environment in which the designed object is intended to exist. A well designed object has an air of inevitability about it, an integrity which can be felt and it fits easily and naturally into its environment. It also brings out and sometimes creates certain desirable features in its environment which would otherwise either not exist or not be noticed.

Design is therefore very obviously an expression in a concrete form of certain ideas and ideals. It is intelligence, intellect and ideology fused into one and made visible. It is important, though, not to forget that by putting ideas into tangible forms the ideas themselves have not been finally and definitively dealt with, that they often continue to live and to grow, and that, in any case, it is difficult to embody a number of ideas and ideals in one visually concrete work. The design of a generation can therefore be considered to be not only an integral part of the physical world from which it sprang and for which it was intended; it is also a pretty reliable mirror of the mind of its age and reflects the attitudes of the age to the world of ideas and ideals in general. Indeed it is impossible to look at the design of a period in perspective without relating it to these factors.

When we do this with today's textiles (having chosen textiles for personal as well as for historic and symbolic reasons), we are struck by a number of things: their tattiness and shabbiness; their lack of cohesion – the jumble of anything with everything – a thoroughly artificial and aesthetically pointless striving to imitate the past, or please the 'working man', or emulate the cheap and the common; their anxiety to conform while pretending to strike out for individuality; their lack of colour and their drabness. It is not that only muddy, uninteresting colours are being used, but that there is a complete lack of control over the colours used: garish colours thrown together anyhow which produce the most unpleasant yet strangely dull, teeth-grinding drabness.

Above all we are struck by the decline in the industry in itself and in its relative importance to other new industries, by the sharp decline in exports. Tracing the history of the industry backwards, decade by decade, comparing achievements, we find ourselves, pretty quickly, deep in the Industrial Revolution, symbolised by the Spinning Jenny, which made textiles so important a part of the economy of this country. We cannot therefore talk of textiles or of industrial design as a whole, of production, sale and design, without talking of the Industrial Revolution. It seems to me that the very success of the Industrial Revolution has in the end become an important factor in today's difficult industrial situation for the country as a whole because it debased the tastes and the ambitions of the majority of the people of this country to an extent which few seem to realise. Since our industry today caters willingly for this debased taste, in most of its products, it cannot compete with the industries of those countries whose aim is to satisfy much better taste. Disappointment however in our performance today becomes meaningful only because it is related to earlier achievement. Our fears and expectations of tomorrow, of being part of Europe associated with ambitious, powerful, pushing trading partners, may be strongly influenced by our understanding of the rise and the decline of British textiles and of the position of textiles in the Industrial Revolution. An insight into the effects of the Industrial Revolution on our lives today could therefore be quite helpful.

In the words of E. J. Hobsbawm in his introduction to his book *Industry and Empire*:

> The Industrial Revolution marks the most fundamental transformation of human life in the history of the world recorded in written documents. For a brief period it coincided with the history of a single country: Great Britain. An entire world economy has been built on or rather around Britain and this country temporarily therefore rose into a position of great influence and power unparalleled by any state of its relative size before or since and unlikely to be paralleled by any state in the foreseeable future. When other countries industrialised this influence and power ended automatically. For most of the world the British era of industrialisation was merely a phase – the initial early phase – of contemporary history. For Britain it was much more than this. We have been profoundly marked by the experience of our economic and social pioneering and remain marked by it to this day.

In his book, Hobsbawm comes to the following conclusions:

1. Economic life in advanced countries has followed fifty-year rhythms (twenty-five up and twenty-five down) and as we have been in the upswing since 1945 we must now expect a downswing.
2. The history of the world over the past five hundred years has been that of the political and economic rise and fall and of its domination by one or another group, based on Western Europe or on European settlers. Within this general development the history of Britain is that of the first phase of industrialisation. Its triumph was that of the pioneer of this phase of history.
3. During and for some time after the Industrial Revolution Britain always had a line of retreat open when the challenge of other economies became too pressing – the Empire, this country's function as the hub of the world's trading, shipping and finance. We did not have to compete and could evade the problems of

competition. Our ability to evade helped to perpetuate the archaic and increasingly obsolete industrial and social structure of the pioneer age. As a result we have for some time now been in decline.

4. Britain has not been able to adapt herself to the new situation where other nations have outstripped her by far. Until she does, Britain must be analysed in terms of decline. True, Britain differs from other countries in that –

(a) She has remained, to her advantage, the most urbanised, industrialised, proletarianised country in Europe. There are no peasants or artisans (nor a large class of small shopkeepers).
(b) She has a two-class system and her working class plays an unusually large rôle in her politics.*

5. But the differences in general and the visual differences in particular tend to diminish. The visual junk of modern civilisation (of airports, of city centres and of beaches) is pretty uniform all over the world.

6. Britain has become more like the other Western industrialised countries but her impact on them is diminishing. The country which gave the world her machines, ships, railways, entrepreneurs, skilled technicians, traders, bankers; the cradle of the concept of sport; which gave the world her basic style of formal and semi-formal male clothing with the special reputation of the craftsman in the West End of London; this country is not doing any of these things any longer. On the contrary we are at the receiving end of other people's influences.

All these points could perhaps be paraphrased by saying that Britain suffers from the Silver Spoon Syndrome complicated by hallucinations. The Silver Spoon of the Industrial Revolution, with which the generations following it were born, has turned into dross. Not realising this, we live in a world of phantasy and with a number of self-imposed handicaps which prevent us from competing on equal terms with other industrial nations.

Is this inevitable? Was the decline inevitable and if not how could the consequences have been prevented? It seems to me that in answering these questions we may understand the past better and also perhaps point to a solution in the future.

*This two-class system has been working to our disadvantage in recent strikes when half the country seemed to be opposing the other half.

There is one strange phenomenon about the Industrial Revolution, at least as far as textiles and most of the old industries were concerned: the machinery was revolutionised but the product remained the same. The Industrial Revolution was a revolution in methods of production, but it was not a Design Revolution. It meant that the same stuff, or nearly the same stuff, was being produced at a quicker rate. Generally speaking, more meant worse; taste became debased in everything and especially in textiles: the Spinning Jenny could produce most ordinary, plain, uniform yarns but nothing with individuality; it could produce almost what was produced before but more quickly. The power loom could produce plain serges but it was quite a while before six-box looms* came into universal use. Again, cloths almost similar to established, hand-woven cloths could be woven more quickly than before; and even six-box looms when they did come along did not radically alter the appearance of what was being produced in spite of the advantages that these looms offered compared to handlooms. The appearance of fabrics has, in fact, barely changed these last two hundred years and one's great-grandfathers, participants in the Industrial Revolution, would find few surprising or different cloths on us today. One is, of course, bearing in mind man-made fibres and knitted constructions, both of which are innovations and both of which have been made to look exactly as if they were nothing of the kind – so near to traditional fabrics do they come in appearance.

This photograph illustrates the contrast between the textile industry and the communications industries in certain parts of the world. The clothes and cloth have not changed for a thousand years; the television set is an up-to-date product of a science-based industry. Even in Europe, where admittedly the time-lag is not so pronounced, it is accepted that people should fly in the latest jets but wear garments (and bowler hats) designed more than a hundred years ago for appearance rather than comfort and practicality.

*The number of colours in a fabric is determined by the number of 'boxes' in the loom since each box can hold a shuttle and bobbin which can be of a different colour. The arrival of six-box looms opened up new possibilities in mechanical cloth production since, until then, only one or two colours could be woven into the weft of a cloth but with the six-box looms up to eight colours could be inserted and thus much more colourful effects and complicated patterns could be achieved.

Textiles and television: a thousand years apart. (*The Times*)

It is clear that as far as the appearance of fabrics, shoes, china and many other objects in daily use is concerned, there has been, so far, no Design Revolution in Britain or even much of a change; and when we compare our efforts in this field to those of a hundred or fifty or twenty-five years ago, we are not admiring the wonderful new ideas in cloths and in fabric design which then suddenly came into vogue and of which millions of yards were, as a result, being sold all over the world. We are admiring and I suppose we should like to emulate, the Victorians' and Edwardians' ability to sell millions of yards of perfectly ordinary cotton and wool cloth all over the world, but especially to the colonies and the dominions. It seems strange – to put it at its mildest – that so much effort should have gone then and still goes today into the speeding up of production and so little into improving and changing the appearance of a product which more than almost any other justifies its existence by its appearance.

Before turning our attention to some of the positive factors which helped the sale of British textiles and other products abroad, it may be as well to examine some of the effects on industry in this country of the British involvement in the Empire.

It has already been noted by Hobsbawm that the Empire served as one of the boltholes to the industry of this country in times of trouble. It is also just possible that the long-term effect of governing a number of backward countries and peoples caused the British to become, to some extent, more like the people they tried to govern than the advanced nations of the world of whom they saw little, with whom they did little business, and from whose competition they sheltered behind the walls of Empire preference and other similar devices. In biblical times it was said to be morally preferable to be, as it were, the tail of a lion than the head of a fox. It was held in those days that an association with the lion, in however lowly a capacity, was bound to be more beneficial to the individual than an association with the fox, in however exalted a position. One subsequently wonders whether even the best colonial administrator, for instance, could forever avoid privately comparing his own standards with the standards of those whom he administered. And one eventually asks whether such a constant comparison, spread over a number of years, was likely to result in the heightening of that administrator's own standards – or in their lowering. No one who has discussed with Britons some weakness or other in the British system in any sphere, can fail to notice that the ultimate reply, when all else fails, and which seems to come so easily to all lips is: 'Ah yes, but look at Russia or Uganda. Isn't their system even worse?' And there are always, say, the Zulus to fall back upon who are bound to be rather worse organised in all respects than we are. The truth is that no progressive country can afford to compare herself with a backward country, just as no athlete, training to win, can afford to take note of any athletes who are not better than he is. The only significant and worthwhile comparisons are with systems which are better than one's own; comparisons in the other direction lead to self-satisfaction, sloth and regression.

In the short and in the intermediate terms the possession of the Empire may have been beneficial to Britain, even if its acquisition was absent-minded and, according to some, lucky, its administration well intentioned from the natives' point of view and its eventual dissolution gracefully accepted and even mildly encouraged. (There are those who, like Correlli Barnett, in his book *The Collapse of British Power*, challenge the assumption that the British Empire was a source of strength to Britain at any time.) But in the long run it has probably made Britons a nation of self-satisfied bit-players. It may have convinced the majority that Britons can learn nothing from anybody and that, on the contrary, most of the things we do are better done than the things other people do. Even when Britons are too polite to say so they seem to believe it with a simple faith. This could be one of the long-term aspects of the colonial aftermath.

Malcolm Muggeridge sums up graphically and as usual exaggerates, the colonial experience in his autobiography, *Chronicles of Wasted Time*. A young man, he is on his way to India by boat, in the twenties; and these are his observations on part of the journey:

As the voyage proceeded, I noticed a strange transformation in my fellow-passengers. They had come aboard as more or less ordinary middle or lower-middle class English, with perhaps, in the second class, the latter preponderating. Now they were changing, the men becoming more assertive, the ladies more la-di-da; as it were, moving farther and farther away from Bournemouth and Bexhill and nearer and nearer to Memsahib status and invitations to Government House. Luncheon was tiffin, a whisky and soda a chota peg, and instead of calling for a steward, the cry was 'Boy'. By Port Said the change was complete.

From the moment of landing in Colombo, I was made conscious of my status as a Sahib. It was like suddenly inheriting a peerage and being addressed as My Lord. Just by virtue of being English and white, if you went to buy a ticket at a railway station, people made way for you. Similarly, in a shop. It was very insidious. At first I found it embarrassing and distasteful; then, though I continued to ridicule it, I came to count upon receiving special treatment. Finally, when for some reason it was not accorded, there was an impulse to become sulky and irritated. From that it is but a small step to shouting and insisting, as, in the days of the Raj, I saw often enough. Our position in India as a ruling race corrupted all concerned; soldiers – and other ranks just as much as officers, if not more so – missionaries, government officials, planters, business men, wives and children; everyone. It also corrupted the Indians, whether they kissed the rod and accepted a position of subservience, or whether they rebelled against it.

It is difficult to disagree with the implications of what he says or to ignore it as one of the possible explanations for the otherwise puzzling inability of the British to relate realistically and fully to most of the rest of the developed world. Today we still have ample opportunity for withdrawing into a whole assortment of mental funk holes with the signs above them showing a surprising variety of colours from the red of the extreme left to the blue of the far right.

Looking at certain positive but hidden factors in the sale of British textiles and other products for about a hundred years (say, from 1850 to 1950) it seems to me that some of them are these:

1. Britain enjoyed and the rest of the world admired (this means envied – in the nicest possible way) the highest standard of living known up to then. Britain also enjoyed more political stability, more individual freedom than any other country and she was militarily the strongest nation on earth. These attributes were also admired. From the superstitious, who worked their magic with an object belonging to their adversary, to the sophisticated who knew that imitation was the sincerest form of flattery, it was clear that one of the quick, most immediate, ways of giving expression to this compound of admiration and envy was to acquire something distinctly British and that this, for obvious reasons, could be achieved through clothes with greater ease than with almost anything else, for 'clothes make the man'. It is no exaggeration to say that people bought British cloth, clothes and other products because they admired the economic, political and military position of Britain. The only means they had of becoming Britons in spirit if not in reality was to dress like Britons.

2. Britain and its textile industry was, in addition, lucky in its squirearchy; much better known in those days and much more influential abroad than the British worker. The British country gentleman set certain standards in the quality of living in general and in clothing in particular which were followed throughout the world and this naturally meant the use of British cloth whenever possible (just as the use of a cloth in a couture collection today can mean its use throughout the world by makers-up in all price brackets).

3. It was mainly in men's clothes that Britain set world standards. The reason for this was the clear relationship in this country between clothes and the men who wore them and the general stability which this set relationship created. One knew exactly where one stood vis-à-vis the British male; one could judge at a glance whether he worked in the city, in a shop or in a factory. Even today it is not too difficult, for the tolerably knowledgeable, to tell not only a person's occupation by his clothes but also his

school and therefore his father's income; his club and therefore his own income; and of course his regiment. There has always been, all over the world, a steady demand for greater stability and probably there always will be. If a whole system of stability can be transplanted by the donning of a bowler hat, stiff collar and regimental tie, then people considered it convenient, pleasant and cheap at the price. (The same goes for systems of instability. How else can one account for the streakers and for the uniform of poverty of the younger trendies today – their artificially faded, frayed and patched jeans, their matted sheepskins, beards and hair.) Clothes were and still are used as social symbols rather than practical objects for wear or protection from the elements. British clothes gave people a sense of security in another way, too. Where else but in a British country house could one spend a formal Sunday, changing clothes three or four times, being absolutely certain each time that one was wearing the right thing? Doing your own thing was not nearly so popular in Victorian times as was doing the right thing.

All the time it was the clothes rather than the cloth which were the catalysts for the sale of British products. The quality of the cloth was good but visually unimaginative and if it was distinct, say a Prince of Wales type of check, then it became rather unsubtle and often completely overwhelming. Essentially then, with every yard of cloth that Britain exported, she gave, as far as the customer was concerned, something intangible of herself which was well worth having. This is what is known in today's advertising jargon as 'selling the sizzle, not the steak'.

4. It is impossible to assess exactly and perhaps to overestimate the importance of these intangible psychological factors of a general sympathy with and admiration of Britain and what she stood for. Only a native of the continent of Europe, who has heard and seen these abstract qualities of Britain discussed at table or on other intimate occasions and translated into action in everyday business life and in everyday shopping, can appreciate their economic importance; and, for what it is worth, I, as a native continental European, now a genuine pseudo-Scot, am relating them to this subject as a whole.

Imperceptibly, however, the situation changed over the past fifty years and Britain's relative position vis-à-vis the rest of the

Prince Charles takes the salute in Hyde Park, Spring 1974. Clothes as symbols (of conformity) rather than practical objects allowed to evolve in response to contemporary human needs. (*Daily Telegraph*)

Clothes as symbols of the dignity of physical work caricatured by the 'anything goes' attitude. (*S. B. Davie*)

A sickness for the past (nostalgia) expressed in clothes which reject the present. (*Bill Coward*)

The lack of clothes as a symbol rather than a practical objective. A 'streaker' at a football match, 1974. (*Daily Mirror*)

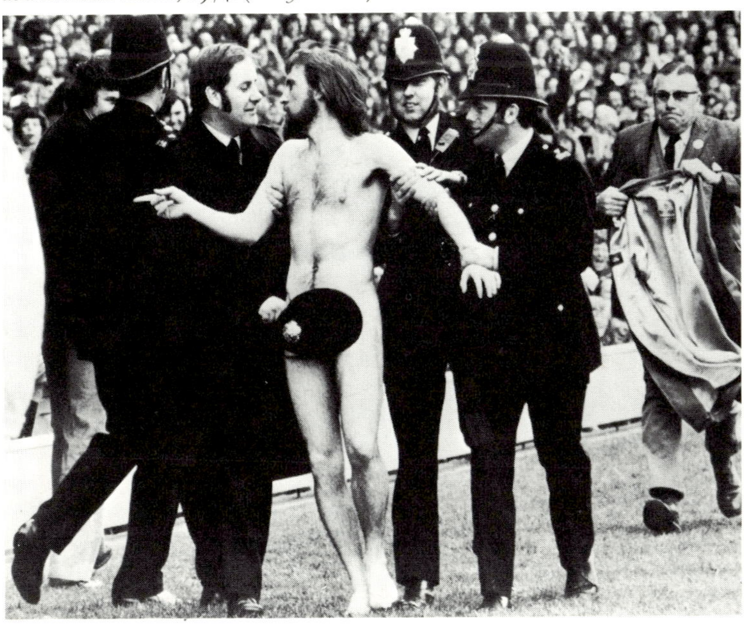

world has changed completely. The world is realising that there is no 'steak', that the real Britain is different from the Britain of their own and perhaps even of the British imagination. The gentle, fecund and sheltering cloud of the Industrial Revolution, which at one time seemed anchored to this island, slowly lifted and Britain became and remains exposed to the hard glare of competition, to wars she could ill afford, to the physical and mental waste and to the residue of her own industrial revolution. Once upon a time where there was muck there used to be brass and then there was just muck. The striving, the explosion of energy which made a manufacturing class out of artisans and merchants, had spent itself. The best brains did not continue to be part of industry and those who did go into it could not be compared favourably with their Continental and American opposite numbers.

Other countries caught up with Britain, founded their own industries and, as nationalism established itself as the new religion and as each country strove to be fully self-sufficient, importing became both uneconomic and unpatriotic. It also became unnecessary. Why import something which can be perfectly well made at home? Simple, standardised, ordinary things can be made even if one is just setting up a new industry. In other words, Britain gradually started to pay the price of concentrating on production and neglecting product design. Today the country and the textile industry in particular, are paying an ever increasing price for taking the easy way out of a classic industrial dilemma. For the strange fact is that almost anyone can follow in the steps of the production pioneer; but it is quite a different matter when it is a question of creative work in aesthetics, technology or in science. Here, well-established cultural traditions are indispensable and cannot be suddenly acquired. By taking the easy way out, by failing to bring its rich heritage to bear on its industrial development, Britain made it easier for others to follow her and to overtake her.

III

In some European countries things were and are different and Scandinavia is a good example. Compare, with realities in Britain, some of the impressions of Sweden as reported in *The Observer* in April 1974 by Michael Frayn, a journalist not normally given to deep analysis of colour schemes and their coordination in interiors:

My friend in Tyresco showed me round a new church youth club. It could have been the headquarters of some successful, progressive advertising agency – suites of conference rooms and offices, each with its own carefully chosen warm colour scheme, its own complete set of matching furniture and equipment.

She also showed me the school where she taught; another warmly stylish new building. Here there were children, but only two hundred and fifty of them in a building which in Britain would have housed two or three times that number. They were taught mostly in groups of not more than fifteen, in warm-coloured classrooms with fitted carpets and house phones. The fitted carpets were clean; when the discreet chimes sounded at the end of a lesson and the pupils moved out of a room, a team of cleaners moved in. The phones had not been vandalised. Nothing had been vandalised.

Foreigners misconceive Swedish domestic style. They have a picture of pale, bland, functional objects made of birchwood and stainless steel against a background of austere white emptiness. But these rooms glow. Their keynote is solid masses of warm colour – deep reds and the dark, flowered greens of traditional painted country furniture. Silver and steel are reflected in dark veneers. Bowls of tulips stand beside blue armchairs. Climbing plants cover walls and half-obscure windows.

I couldn't help feeling, as I watched one young drunk with blond hair over his shoulders aggressively bum a light for his cigarette, then struggle halfway up the platform, against some powerful magnetic field which seemed to be trying to divert him from his goal, in order to dutifully deposit the empty cigarette packet in a rubbish-bin, that there was something rather homely about even these deviations from decorum.

This air of changeless prosperity is deceptive, for the whole experience

"Conformist!" 16.vii.71

"Excuse my asking, but which side are you on—pollutionwise?" 12.v.71

From *Theatre in the Flat* by Osbert Lancaster (John Murray).

Pop festival, Britain 1974. (*Daily Telegraph*)

of modern Sweden has been one of continuous and dislocatingly rapid change. God did not create Sweden rich and progressive. Until far on into the nineteenth century it was one of the very poorest and most backward countries in Europe.

In Scandinavia, it would seem, there was no shifting of the rural population into cities to anything like the same scale as in Britain. People were, therefore, not wrenched away from the enjoyment of ordinary things in everyday life, carefully made, constantly used and lovingly preserved. Their fingers never lost the feel of the hand-shaped handle of the axe; they knew the wood-and-leather sound of the homing handloom shuttle and the touch of homespun fabrics. Their enjoyment of and desire for patiently and lovingly made objects never atrophied, their eyes never lost their sensibility to their environment. There was no sudden world demand for the products of these countries so there was no need for the building of vast slums around the factory-prisons as in this country. Most Scandinavians continued to lead natural lives and went on using their eyes and their senses normally in a pre-industrial revolution, unspoilt environment.

In addition, even fifty years ago far-sighted educational efforts were being made in these countries to inculcate an awareness for good craftsmanship and design so that to surround oneself with well designed objects in the home became a genuine need and a way of life of the average citizen rather than an act of conformity to the tastes of one's neighbours. The ritual whitewashing of the front step in the terraced houses of industrial Britain, the 'shepherd boy and dog' in most working-class windows and the aspidistra in the middle-class ones, the 'flight of geese' on the walls of both classes are examples of the kind of mentality where lives are spent conforming rather than fulfilling genuine needs. Their modern equivalents are perhaps the ritual polishing of the car (perhaps the car itself), the joining of the endless weekend queues to the congested seaside and the disproportionate emphasis on and expense of the annual holiday when compared to the standard of living in general throughout the year. (The urge to conform is probably not less in Scandinavia than here but the Scandinavians have somehow managed to fuse their desire to conform with other genuinely felt, rather than with artificial, desires.)

Then too, the Trade Unionist movements developed differently there than they did in Britain. Perhaps more slowly, subserviently, Germanically (obediently?) if you like, in the sense that they were prepared to bow to authority, to the 'master', perhaps with less insistence on 'absolute freedom' (to be bloody minded?) but also perhaps with more concern for the trades they represented than their British counterparts who were interested in power for its own sake. For if there is a basic difference between the craftsman, the businessman and the Trade Unionist, it is perhaps that the craftsman will give something in return for nothing, the businessman will try to get something in return for something, and the Trade Unionist will insist on getting everything and on giving nothing. The craftsman stands alone and does what he feels he has to do and the reward is not the only, or often even the main consideration; being a craftsman means that he enjoys his work and that it, in itself, is part of his reward. The businessman works with his capital, is more or less organised and does what he feels he can to defend his own interests; few businessmen would be in business if they did not have to be. They work so as to free themselves from work. By the same token they realise that in order to remain in business they must offer some value for money. The Trade Unionist by contrast is convinced that his existence is justified in absolute, socio-philosophical terms, is completely organised, has nationwide, often world-wide monopoly on labour, feels himself therefore to be outside the law of give and take, and does what he wants to do regardless of the interests of other groups in the community – often regardless of the interest of the state itself.

British Trade Unions have more power than Trade Unions in other countries today: no one dares to challenge their basic assumptions. They have, in Britain, been accorded a position similar to that of the church in the Middle Ages. Yet it should be as natural as it is necessary that the community which dared to question the godly origins of Man should dare to doubt the right to absolute power for which the Unions are reaching.

It is important to question the moral case for and to examine the practical effects on the quality of industrial products of an organisation whose main aim is to ingrain into people a general attitude towards work which, instead of seeing it as a natural outlet for a number of inborn drives and talents, depicts it as an instrument of oppression in the hands of the exploiters; from which it follows that work should be done grudgingly; that it is one's duty to do as little of it as possible and to extract for it, in payment, as much as possible; which justifies, legalises the use of, threats, bullying, coercion, blackmail and violence (plain words for 'demanding substantial increases or facing the possibility of industrial action', 'expulsion', 'closed-shop', 'strikes in sympathy' and 'picketing').

Consider the beneficial effects on the European and especially on the German economy of the ample availability of labour, of the alternative to the 'I know my rights' worker – of the 'guest worker' – and of this counterbalance to the power of the Unions. Try to imagine the reaction of British Trade Unions to the suggestion of bringing, systematically, three million European workers into Britain. (They are quite prepared to accept two million Asians under a variety of post-imperialist slogans, unsystematically, and to muddle along with them now; but then muddling along and not facing issues has almost become our way of life.)

This is how Theo Sommer described (*Encounter*, November 1973) the position in Germany:

Initially West Germany had ten million deportees from the East and then three or four million refugees from the Russian zone of occupation. For the first fifteen post-War years this provided a full reservoir of labour for the 'economic miracle' – the 'Wirtschaftswunder'. The massive influx of foreign workers began when the late Walter Ulbricht made the frontier between the two German states watertight by building the Berlin Wall. In 1960 there were only 300,000 'guest workers'; by 1965 there were over a million and by 1971 over two million. Since then the figure for legal foreign immigrant workers has risen to 2·4 million and by the end of 1973 it will be between 2·5 and 3 million.

Our Unions were prepared – up to a point – to accept the unrealistic relationship between the Commonwealth and Britain and to allow workers from the 'Commonwealth' to come here. But until we went into Europe there was no question of letting Europeans come here, even though by tradition and background Europeans have much more in common with Britons than Asians and Africans do and would therefore have settled more easily in this country than Commonwealth citizens. A fiction seems to be all right with us as long as it fits in with accepted mythology; facts are unacceptable if they involve a rethinking of attitudes.

The influence of British Trade Unionism, with its exaggerated emphasis on and its idiosyncratic conception of 'freedom', on the products of our factories, must therefore be profound.

In certain European countries where crafts were kept alive in spite of industrialisation the attitude to craftsmanship spilled over into industrial activity, the factory worker transferred a good deal of his attitude to his craft to his work in the factory. He remained primarily interested in doing a good job because he could not help his nature. By doing this, by making good things (and so giving good value) and by demanding good things as a customer, he helped his country in the long run. A good craftsman appreciates the work of other good craftsmen and has a positive attitude to the artefacts in daily life. A country whose workers are craftsmen automatically attains a better quality of life and therefore of consumption than a country whose population has been pauperised and deprived of any feelings for quality in the little things which determine the tenor of everyday living. The workers' and the average man's attitude to craftsmanship, design and quality will in the long run influence the nature of demand and supply in any given country, especially in a country like Britain where there are proportionately more wage-earners than in any other country. The nature of this demand, its character and aesthetic level have been and still are consistently higher in Scandinavia than in Britain.

In addition to the reasons we have just looked at, this can be explained partly by the fact that, in industrial Britain, the philosophy of getting something for nothing is shrugged off and accepted; whether out of cynicism, lethargy, helplessness doesn't really make any difference. Perhaps the following extracts from 'The Lion and the Ostrich', a talk by Arthur Koestler delivered to the British Academy in 1973, may help to cast some light on these British attitudes, its effects on industry and on Europeans who observe them. After narrating some of his early wartime experiences as a soldier in Britain, Koestler went on to say:

My company – the 251st Company Aliens Pioneer Corps – was engaged on a fairly vital defence job to protect the petrol reservoirs in the vicinity of Bristol. (We were digging craters which, during air-raids, were filled with inflammable liquid and set ablaze to convince the raiders that they had accomplished their mission and could safely go home.) We were glad to do a useful job, but as you would expect from aliens, we became over-enthusiastic, so we asked our Commanding Officer (who was British) to do away with the ritual tea-breaks which – what with downing tools, marching to the distant cook-hut and back, morning and afternoon – cost nearly two hours of working time, in addition to the lunch-break. The C.O. expressed his appreciation of our laudable zeal and explained that we had to have our tea-breaks whether we liked it or not because the British Pioneer Companies and the local unions of our civilian workmates would raise hell if we didn't. The time was about nine months after Dunkirk.

Most of our cosmopolitan bunch in the 251st Company came across similar experiences, which delighted us; they made us feel that such holy innocence had an unconquerable quality. But other experiences left us rather bewildered. While digging for victory, we came into intimate contact with working-class life and found it fundamentally different from its Continental equivalents. In the NAAFI canteens, in the pub and later at the snooker table in a London ambulance station, I was taught to accept the stubborn persistence of the hoary cliché that people in general were divided into Them and Us.

But that 'Us' had nothing to do with class-consciousness in the Marxist sense, as it existed in the Socialist and Communist Parties of Europe. Marxist dialectics was as much double-Dutch to the British working-class as it was to the rest of the nation. Instead of the fierce class-hatred which had scorched the Continent with revolutions and civil wars, there was an almost smug acceptance of living in a divided world, as licensed premises are divided into saloon bar and public bar. On the Continent, the symbolic gesture of militancy was the clenched fist;

here it was closer to a shrug, a deliberate turning of one's back on middle-class standards of values, codes of behaviour, vocabulary and accent. Off duty our workmates were lively characters, full of fun and games; on the work site they moved like figures in a slow-motion film, or deep-sea divers on the ocean bed. They seemed to be conforming to a sacred doctrine, a set of unwritten maxims of life; go slow and take it easy or you are letting your mates down and we shall all be on the dole. It's a mug's game, anyway, and you are in it for life unless you hit the pools.

In the Libyan desert, or as rear-gunners in a bomber, they would have done a magnificent job; for in those circumstances the gulf would have been temporarily bridged by shared danger and hardship – and by the awareness of playing a man's game instead of a mug's game. The same lovable bloke who risked his life on D-Day to keep the country free would not lift a finger at Dagenham to put the country back on its feet. The British working class seems to have become a powerful non-competitive enclave in a competitive world. To appreciate the contrast, compare John Braine's *Room At The Top* with Alan Sillitoe's *Saturday Night And Sunday Morning.*

During the early post-War period in Italy and France the Communist Party was the strongest single force both in the Trade Unions and in Parliament and both countries seemed on the verge of civil war. But the rising curve of prosperity led to a corresponding decline in revolutionary fervour. Moreover, on the Continent there exist Socialist, Communist and Christian unions which compete for the workers' favours; and their openly declared political programmes are massively represented in Parliament, so that Trade Union politics has become a truly democratic game. In this country events seem to have moved in the opposite direction. Militancy in the Unions has increased instead of decreasing with growing prosperity, with a tendency to harden into the kind of cold class war where passive majorities are led by active minorities dedicated to idealogies which cannot muster a single elected representative in Parliament. This is a strangely paradoxical state of affairs in the oldest democracy in the world; it contributes to the industrial malaise and plays a significant part in its poor showing in the European league. But it is rarely ventilated in public debate. An ostrich-like attitude is considered more appropriate, particularly by progressive middle-class intellectuals haunted by guilt

One hesitates to add to an already – and, to my mind, quite unjustifiably – overburdened national conscience yet another reason for brooding over an imagined or real wrong. It is these groundless, national neuroses (towards Ireland, towards a number of countries in Asia, towards many more in Africa; you name it and the British will admit to having done something nasty to it and to being very sorry about it) which prevent us from dealing with many of our urgent, current problems naturally and effectively. Yet one feels, when reflecting on Koestler's penetrating comments, that his lack of peacetime industrial experience has perhaps caused him to miss an important and fundamental aspect of industrial life and its influence on workers' attitudes in Britain. The opening scenes of the film made of *Saturday Night And Sunday Morning* show, with great veracity, the engineering workshop where the 'hero' of the story works. It is visually a mean, man-made but inhuman desert of devastating impact on self-respect with absolutely no concessions to human feelings or to personal dignity, and is unfortunately quite typical of most British factories and workshops. If one is condemned to spend most of one's days in such an environment, what other choice is there apart from sleepwalking?

The British workers' attitude must, to some extent at any rate, have been moulded by the British employers' attitude to the worker and to working conditions. I have the impression, without being able to prove it, that in this respect the attitude of employers in many Western countries is and has always been in advance of British attitudes. I suspect that human dignity is bolstered and nurtured much more effectively by small, daily attentions to the almost insignificant-seeming, personal requirements than by the occasional weighty, cerebral, politico-philosophical pronouncements on the condition of humanity.

In the words of R. H. Tawney in *Religion And The Rise of Capitalism*:

> The most obvious facts are the most easily forgotten. Both the existing economic order and too many of the projects advanced for reconstructing it, break down through their neglect of the truism that, since even quite common men have souls, no increase in material wealth will compensate them for arrangements which insult their self respect and impair their freedom. A reasonable estimate of economic organisation must allow for the fact that, unless industry is to be paralysed by recurrent revolts on the part of outraged human nature, it must satisfy criteria which are not purely economic.

Whatever the reasons, there is in Britain, on most social levels, a diametrically opposed stance to that in many other countries, not only to giving value for money in the round but to the very idea of craftsmanship (which means a total identification with the product), of caring for abstractions in the visual and tactile fields and in actually satisfying the demands of the senses. This British attitude, percolating from the upper social strata and from the managerial class, opposes the concept of design as part of the industrial process, identifies it as being wasteful, effeminate and unprofitable and makes the possibility of a design evolution in Britain rather unlikely and certainly difficult. It quite simply ignores most of the environment as a personal or industrial factor.

Designing is nearer to craft than to business. It is certainly not an exercise in a struggle for power with which one is familiar in big and small businesses. Nor is it the satisfaction of greed by other means. It is a 'give' rather than a 'take' activity. It will flourish in conditions similar to those in which craft will flourish; and just as little worthwhile will be produced by the craftsman who constantly asks himself 'how little can I get away with', so designing cannot flourish if the accent of the organisation for whom the designing is done is on 'taking' rather than on 'giving and taking'.

There are other factors, in addition to the British kind of Trade Unionism and to the visual insensibilities of British management, which have combined to create a situation in this country where no long-term thinking in regard to the design content of industrial products has been or is being undertaken and some of them are these:

Over the past one hundred years and not counting the past thirty years there have been major movements in the visual arts in certain Continental countries. The Impressionists, Post-Impressionists and Cubists in France; the Blue Rider, the Bridge, the Bauhaus in Germany; the Abstract Expressionists of the post-war period in a number of countries. By their very nature it took several decades for these new ways of looking at the world to be understood and accepted even by those whose interests lay in painting and sculpture. It took even longer for them to percolate into the consciousness of the man in the street. When it did happen it probably influenced not only the accepted way of looking at works of art; it meant that the man in the street had equipped himself with completely new tools for looking at his environment and the objects in that environment. People began, literally, to *see* things differently. The standard of demand for the things which comprise everyday living in Britain, where there were no similar important movements in art, remained stationary or, as a result of fads, of which Carnaby Street is a recent example, it went backwards. Serious efforts to improve design standards in Britain have been made by William Morris; by Eric Gill; by Sir Francis Meynell (particularly by persuading the Government to improve the quality of their printing, through HMSO); by the team of designers, chosen by Sir Gerald Barry, who created the Festival of Britain in 1951; by the Council of Industrial Design, initially created in 1915 and then called the Design and Industries Association; by Sir Robin Darwin and his reform of the Royal College of Art; and by the drastic re-shaping of design and art schools in England and Wales culminating in the National Diplomas in Art and Design, now in process of being converted to CNAA Degrees. But, in my opinion, not one of these men – except William Morris – nor any of the bodies or institutions, has succeeded so far in influencing the ordinary man in this country. The quality of life in general and of things seen in particular has, for those who live above the aesthetic subsistence level, by and large, deteriorated steadily and continues to slide on a seemingly endless downward slope. We have more of everything for more people but what we have is getting steadily worse rather than better.

I am saying this fully conscious of the continuing existence of the Design Council of whose Scottish Committee I was a member for eight years. The Council seems to me to have had almost no influence on Art Education at any level nor does it appear to recognise that this is an area of primary importance in its sphere of activity. It is difficult to be accurate in assessing the Council's influence on industrial design. It has displays of British industrial products selected by quality, but, since it derives a considerable part of its income from firms whose products it displays, its influence on the design of these products would seem to be little greater than the influence of newspapers on the design and the content of the advertisements which appear in them. I doubt even if the Council can justify its existence simply as a Complete Cataloguer of current British industrial products. It can certainly lay no claim to being a significant contributor to the shaping of design trends in Britain, to uncompromising encouragement of uncompromising good design or to the clear-cut discouragement of bad design. In fact the Council seems to encourage the trendy and to fail to encourage good design when to do so means being out of step with the latest fads and fashions.

Britain as I have known it since the war has been a country where the standard of design for home consumption has been neglected and rarely has it been properly encouraged to improve. Often it has been and is being perversely misdirected for the gain of short-term advantages. Industry needs a broad base for the sale of its products and this is usually its home market. Rarely can industry produce one thing for the home market and a basically different thing for export. But this is exactly what British industry has to do if it wants to export, since a comparison of the average living standards – and I take this to mean and to include the average design standards – between Britain and Europe, will show our standards to be appreciably lower. It will show that what we use in this country is unacceptable abroad and what we make specially for use abroad is, not surprisingly, not as good as the local product where local industry, working with first-hand knowledge and an instinctive understanding of its public, has the advantage. Nor should we expect designers, whose time is taken up with satisfying an indifferent type of demand at home, who have no opportunity therefore of evolving something original, topical, exciting of their own, suddenly to produce ideas which will leap-frog over both our products for the home market and the more advanced products of Continental manufacture. This is not natural and is asking too much. One does not find one's soul by spending all one's energies in doing what others tell one to do. Working within limitations is good discipline and necessary; but working only within limitations is in the long run a sure method of desensitising the designer, debasing the product and destroying the company.

Another comparison which I intend to make between Britain and European countries is perhaps as important as any. The status of the designer, as reflected by the protection the law gives the designer and his work, is far higher in Scandinavia than it is here. Thus it is the law there that, unless there is a clear agreement to the contrary, a designer is automatically entitled to a five percent royalty on all his work for as long as he lives and that his heirs are entitled to it for fifty years after his death. There are a number of other laws, usages and attitudes, all of which reflect the high value placed by the Scandinavians on the work of their designers; on the protection and the encouragement of those who are mainly concerned with giving from those who are mainly concerned with taking.

Looking to the future, it must be obvious that if what we design is not wanted at all or not wanted in sufficiently large quantities then we will have to change our product; we shall have to produce something which is more attractive to our prospective customers. From producing things, in which we do not believe, merely because we hope to sell them, we shall have to change to making things which are not wholly imposed on us by outside necessity but are also the result of inner needs and convictions; in which we believe and which we can convincingly promote and help to sell. The appeal must be first to the eye and to the senses. There is no doubt in my mind nor in the mind of retailers that when a woman is confronted with, say, one thousand dresses in a shop, she will make a bee-line for the dress of the 'right' colour and fabric design (right for her), then she will consider the style and finally the price; and within surprisingly wide limits she will adjust her ideas on price, but not on colour and design. The same goes for men; and it is true of people when they shop for cars or carpets. It is obvious therefore that the first and most important factor, without which the others do not begin to be considered, is, in textiles and fashions and in most other industries, the look and the appeal of the article, its colour and its design; it is the general impression created in the first few seconds of seeing the product. I should like to suggest that the best way of achieving worthwhile changes is not by copying our competitors, nor by polling our potential customers and then doing what we think that the majority seem to want, but by evolving a genuine and valid point of view of our own which must make sense within a British context and which would be desirable and comparatively advanced aesthetically speaking in present-day terms. This implies two things: a lifting of aesthetic standards of demand in general, a house-putting-in-order operation of industry as a whole; it also implies a change in our approach to design: a realisation that while progress in methods of production is indispensable in a healthy industrial country, to change only the methods of production and to go on producing old-fashioned things on the most up-to-date machinery is self-defeating commercially and destructive spiritually. It is like producing atomic energy to keep a windmill going on days when there is no wind.

We in Britain need to make design a more important factor in the

German views of Britain from *Punch*. Above 'How do they see our architecture?' by Rauch. Below 'How do they see our women?' by Hüsch.

'I always said that all this European emancipation business would destroy the British way of life.'

future than it is now but this will only be valid and effective if our designs come from within us and express our genuine, individual and, one hopes, worthwhile points of view. The importance of approaching design from this angle now that we are in the Common Market can be seen from an interview on the BBC concerned with a totally different industry from textiles – with book publishing.

Sir George Weidenfeld of Weidenfeld & Nicolson had this to say, amongst other things, to Kenneth Harris in this interview in the early part of 1972:

How do British books and writers rate in Europe?

As far as books are concerned, we are already in the Common Market. There are no tariffs, there are no restrictions on the import or export of books or royalties. But with our entry into the Common Market there will be a greater incentive for European and British publishers to work together. Mind you, there are one or two paradoxes in this present situation. In a curious way, though publishing has become more international, there has also been growing up a new kind of cultural chauvinism in various European countries. Ten years ago, if you went to Germany, France or Italy, and said "Here is a book by Kenneth Harris on an important European problem – on the subject of pollution or on the problem of the younger generation", the mere fact that he was a British writer and a contributor to a leading Sunday newspaper, would immediately have made him acceptable to a German publisher, who had a curious sense of inferiority or inadequacy when it came to German authors. Today the situation has completely changed. German television and newspapers have built up national reputations, the German publisher has much greater confidence in his own writers and in his own style of presenting a book.

The same reasoning with greater force applies to industry at large. The mere Britishness of the product doesn't impress customers any longer. Our entry into the Common Market is the formation of a partnership and this implies equality. Europeans do not any longer consider Britain superior in any way to themselves. The collective prestige of the country does not help us commercially any more. Britain is being discovered by the millions who come here every year and she is being seen for what she is, not a country leading, but often behind many others in some and often in many respects.

It was broadly on these lines that the German industrialist and politician, Philip Rosenthal, spoke at a Design Conference in 1972 in London. First he graded rather elegantly, I thought, the rôle and the importance of design and the designer in industry according to the product, and used Turbines, Teapots and Ties for summarising the essential differences. Thus in turbines where, because of practical considerations, the aesthetic design content is of minimal importance, the rôle and position of the designer can also be comparatively unimportant. Teapots have to be functional and nice to look at and here the function of the designer is much more important. In ties, where appearance is all and practical function minimal, the rôle of the designer is very important. He went on to say that design must be an integral part of the product and must not be 'smeared' on to it which brought him to the most telling point he had to make: that even though the German market probably needs special things, it would be just as wrong commercially and wasteful to make, for export, German products in Britain (that is products which are right for Germany and similar to those being produced there) as it would be to make British products in Germany and then import them here; and of course it would be quite wrong to take a British product and 'smear' on to it some characteristic which might make it acceptable in Germany. He thought that what the Germans wanted and needed were typically British things which fitted into and would add something to and enrich their own living patterns. I would interpret this as British products which contained within themselves what was distinct, characteristic and desirable in the British approach to living, which were yet not so diametrically opposed to the German way of living as to make their inclusion into that milieu too jarring.

What we need when we think of exporting is an individual outlook of integrity which will offer something different – and amusing – and sometimes something better than the locally available product. But only if this design point of view – perfectionist, particular or even finicky – is part of a larger more comprehensive point of view towards living and working in general, will it have any real validity and will it lead not only to a richer, fuller life at home but also to better and more profitable exports.

We need to know our own minds; we must attempt to pursue the satisfaction of the genuine human needs of the body and of the senses without shame or inhibition but with integrity and individuality. We can then hope to widen and enrich human experience, improve the quality of our own life and to help others with whom we trade to do the same.

IIIA SCANDINAVIA AS I SEE IT

The pictures which follow are not intended to be taken as a balanced presentation of everything that can be seen in Scandinavia; they are merely a very personal way of summing up part of one designer's feelings about these Northern European countries. To me they mean these things, among others; this approach to everyday living.

Scandinavian factories, seen from the main roads, are satisfying and sometimes quite exciting to look at. Inside they are clean, tidy and appear efficient. Workers can often be seen sitting in the sun around coloured tables, under gay umbrellas among shrubs and flowers, during their midday break; if they prefer to stay indoors they can do so and eat and relax in well-appointed canteens.

Municipal and co-operative housing estates have an up-to-date, attractive appearance, look carefully built and are well sited. They and the surrounding spaces have that cared-for look. Private houses and farm buildings, old and new, are very obviously lovingly looked after.

Hotels, on the whole, are clean and comfortable and not expensive for what they offer and homes have a very special atmosphere of ordered calm, warmth and attention to detail which endows apparently ordinary things with hidden significance and makes the visitor look at, notice and re-assess objects that he would not ordinarily have 'seen' at all.

Most of the things from Scandinavia shown here are both beautiful and useful; in one or two (the cutlery, for instance) practical aspects do not match visual appeal. Scandinavia, like some of the things I have selected to show, is not perfect; but as far as the general appearance of its industrial products is concerned and in its own appearance, it is a very different place from Britain.

I have designed carpets for a factory near Aarhus in Jutland for many years and I know this part of Denmark better than the rest of the country. There are many worthwhile industrial buildings in this area but one which I pass every five or six weeks made a deep impression on me the first time I saw it and is shown here.

Some Danish factories supply not only Scandinavia but many countries in Europe and are therefore quite large. These pictures, taken on an ordinary-looking day and without any prior notice, show the typical tidiness inside one such factory producing tufted carpets.

Lunchtime sun-seekers at the carpet factory.

Others who prefer the shade.

Council housing, Aarhus, 1973.

Cooperative housing, Aarhus, 1973.

Council housing, Aarhus, c. 1960.

Cooperative housing, Aarhus, c. 1960 with minimal openings towards the very busy and noisy main road.

The same estate seen from the inside. The houses have privacy, are quiet and have trees, little gardens and play areas around them.

Swedish glass.

Swedish pottery.

Rosenthal china.

Finnish tableware.

Swedish glass. Shown here upside down in a chandelier arrangement to emphasise light and shape.

Danish cutlery.

Danish pepper and salt mills.

Finnish kettle and ice bucket.

A Danish chair in steel with rope seat and back.

A Danish chair in wood.

The Swedish Stand at the Exhibition of Household Textiles at Olympia, London, October 1974.

The British displays flanking the Swedish Stand at the Exhibition.

IV

The view that it would be wrong to take a British product and smear on to it some characteristics which might make it saleable abroad may make some down-to-earth, practical people exclaim:

> Why not? If this is what sells, as in the case of multiple tailoring, is this such a wrong approach? Fashion is so quickly universal these days that if wide trousers with turn-ups are 'in' here, they are almost simultaneously 'in' in Milan, Marseilles and Mannheim. If, as is currently happening, basic designs are created here and then, with the aid of knowledgeable overseas designers, the required national handwriting of variation in cut, or additional buttons or belts is added for the product intended for those markets, is such sensible action to be dismissed as 'smearing' (horrid word!) with its implications of not really caring or even trying?*

What, in other words, is wrong with taking any well designed product and allowing a number of people to add to it or take away from it if this brings about a degree of acceptability in certain markets and means better sales?

There are, broadly speaking, two answers to this question, each based on a basically different approach to the function of the manufacturer.

When we examine the attitudes of people to business and to industry we find, at the one extreme, those who, like land or commodity speculators, see nothing wrong in exploiting certain given situations entirely to their own advantage regardless of whether their part in the transaction contributes anything of any value to the object of their manipulations. Thus, to take a contemporary example, about ten times more cocoa is sold on paper in the world commodity markets than is grown in all the producing countries. Cocoa is, therefore, many times more expensive than it

*Dr P. W. Smith, President of the BTEA and Chairman of the Shirley Institute, in a letter in the June 1973 issue of *The Textile Design and Institute*.

would otherwise be because it changes hands (on paper) so often; its quality, however, remains totally unchanged by the higher prices; the people who grow it are no better off; but a series of middle men, the buyers, the sellers and the manipulators, benefit hugely.

At the other extreme, we have craftsmen and people in the craft-based industries who, in the process of manufacture, change the raw material out of all recognition and increase its value by adding to it the intangible components of their labour, inborn gifts, acquired skills and often scientific organisation and production methods. In some European languages this kind of activity is described as the 'Ennobling of Raw Materials'. Those who spend their lives in this kind of work consider it, on the whole, immoral to exploit situations to their advantage, to buy and to sell without adding something of value to the goods that they handle. They believe that they have to give before they are entitled to take.

In between these two extremes there is an almost endless spectrum of differing attitudes to work, to business and to industry where the intended value-enhancing element and social-justifiability-content of a product varies from person to person and from company to company.

If, as a manufacturer, one takes up a position nearer in outlook, in relation to one's work, to the first group of people than to the second, then it is easy to see that the question contains its own, immediately obvious answer.

If, on the other hand, one's outlook is somewhere in between the two extremes or nearer to the second than to the first, then there are bound to be constantly recurring difficulties of definition and decision in regard to the quality, amount and the integrity of one's work.

The short answer to the question: 'What is to stop any number of people from changing a product to make it sell better?', in the first case then, it seems to me, is: 'Nothing; not if immediate sales matter more than any other considerations; if this year's profits mean more than the basic attitude to work and to living which a range of products usually reflects.'

There are indeed places where nothing matters more than immediate sales. For instance, in certain parts of America, one learns, one can watch a Frankenstein thriller or a performance of Macbeth on television and one can decide during, say, the fourth commercial break, whether one would like to see a bloody ending or a sugary one. All one has to do is press one of two buttons.

People who are in the entertainment world simply for money behave logically when they supply the public, as far as the law permits, with violence or pornography or Bowdlerised classics, if they feel that that is what the public wants. They have no other criterion by which to judge the failure or success of their activities except by their last balance sheets. So if the public is prepared to pay for a Shakespearean tragedy with a happy ending, but is not prepared to pay for one with a bloody ending, some people consider it the legitimate business of show business to give them just that.

The basic literary product, be it a Frankenstein-type thriller or a Macbeth-type classic, is there. The motivating force in the creation of the original in both is something other than only, or even mainly, money. If these plays have any real value it is derived from and reflects the essence of the personalities and the abilities of the authors. Yet, it is not difficult to imagine the usual conference between producers, directors and financiers when the original idea is considered: 'Why not take the central theme,' the business promoters ask, 'and graft on to it something that an audience wants? Something for which they are prepared to pay? And take away something some of them might object to. It's still Frankenstein or Macbeth, isn't it?'

Well, no; of course it is not.

The assumptions on which I sense the question to be based, conjure up memories of a top executive in a well-known carpet manufacturing company who said, jokingly I thought at first, but repeatedly and as I realised later, quite seriously, that their business was to make 'Muck for the Masses'. This company did then and continues to do today, neither better nor worse than the average British carpet manufacturer; and like the others, it naturally sells by far the largest percentage of its products in this country. (Oh yes, they export; go to a Continental trade fair and you might see them standing there uncluttered by enquirers, splendidly visible for miles with other British carpet manufacturers as their only company; there, amongst acres of antiquated absurdities, they gaze into the distance while the rest of the fair goes like a bomb.

To round off this very practical approach to carpet manufacturing, the 'Muck for the Masses' motto can be supplemented with a simple, ingenious and aptly named 'Char-Test', applied to all designs and to all advertising. If the company's charladies like them, they go ahead. If they don't, it is back to the drawing board for everybody: in some companies' views, the most direct and practical way of giving the public what it wants.

I have seen the Char-Test vary in sophistication; it can just as well be a Typing-Pool-Test (ask the typists what they think) or a Family-Favourite-Test (ask the family what they think). The company policy behind it is less than sophisticated: 'Let's give them what they want first and we'll cry all the way to the Bank afterwards.'

I am no archbishop nor, I hope, a prig; I think that I have, nevertheless, in what I have said so far, made clear my attitude to this kind of approach to working, designing and manufacturing. Anyone who has followed my attempt at the definition of design will have had more than a hint of what my answer to a question put so explicitly is bound to be. Considering Industrial Design, as I do, to be not unlike, say, Architecture and thinking of Architecture as similar in essence to the other Arts such as Painting, Music and Literature, I look for a sense of unity in any work containing Design, just as I look for it in Architecture, Painting, Music and Literature. Whether I like the work as a whole or I do not, I can respect it only if it succeeds in convincing me of its sincerity, consistency and unity. I may not agree with the point of view expressed in it but if it stems from personal integrity which is reflected in unity, it is valid.

I do not necessarily think of creative integrity exclusively in terms of single individuals. True, I know of no painting painted by more than one painter, or of a poem written by a team of poets; but we all accept the validity of the collaboration between the composer and the librettist, between the composer and the choreographer. We accept even more the totality which results from the collaboration between the director and the stage designer; between the man who imagines and the man who makes the thought solidify into tangible, colourful fact. But in these instances there is a deliberate collaboration and this presupposes a basic agreement, that is, unity, between the collaborators who then go out of their way to see each other's point of view and meet each other at least half way.

If, then, one of the hallmarks of a creative work is unity it is obviously impossible to expect the initial unity to be preserved if outsiders are free to introduce changes at will or cause them to be introduced (no matter what the intention behind the change is).

I should have thought that a better way of tackling the problem of the different requirements of different markets was either to brief different designers for each job, choosing the designer to fit the job, or to design articles which are so outstanding that they have a universal appeal and application without the need for adaptation. In the world of art it is accepted that a work remains inviolate even if, as in 99 per cent of the cases, only one artist is responsible for it and even if the work has to be seen or heard all over the world. It would seem to me that we could at least aspire to a similar aim and attitude in Industrial Design.

The trials, tribulations and the bankruptcies in the film industry where everybody tries to please everybody else and strong, original works of literature are mutilated in order not to offend the censor of this country, the Mrs Whitehouses of that country, or the government of yet another country, are too well known to be gone into here. That all these efforts at pleasing everybody have led to the closing down of many studios makes any comment on the commercial effectiveness of such an approach in the long run, superfluous. In another area, the story of the Ford Edsel* has a simple and sharp lesson to teach us about trying to play safe, about trying to please everybody and about ignoring one of the fundamental principles of Design, which is unity.

In parentheses, one of the strangest and at the same time most significant phenomena on the industrial scene may well be the shift in the Ford Motor Company's philosophy from: 'They can have it in any colour as long as it is black' to the 'Let's please everybody with one car' philosophy of the Edsel, and more recently to the 'Design Your Mustang' promotion, which in essence caters for the increasing demand for individuality, and means that Ford is trying to please everybody not with a mass-produced car from their factories but the customer's own, individually designed car. In other words, they have given up trying to find out and trying to make exactly what the customer wants. They supply the components and allow the customer to get on with it and put together his own car.

*Described in *The Fate of the Edsel and Other Business Adventures* by John Brooks (Harper & Row, New York, 1963).

Yet I admit that I do not by any means consider the question of designing without compromise settled for good. I often ponder over the implication of giving the public what it is supposed to want. The more so because I realise that giving the public what it wants is not necessarily or simply an instance of cynically making hay while the sun shines, of exploiting the comparative ignorance and helplessness of people, but a temptation to which most of us succumb in varying degrees one way or another. So, I often ask myself, do we knowingly make inferior products to satisfy an assumed demand – an assumption which rests on past sales records and is therefore pragmatic but correct – or is our conclusion that the public wants inferior products erroneous because it is based on biased and one-sided data? Has the public, in the past, and does it at present, I wonder, choose bad design because it had and has no alternative? Would it have chosen good design if in addition to bad design it could also have seen good design which was not completely outside its reach in price?

This is not a question of educating the public; there is no doubt in my mind that it is not now, nor should it at any time be, the manufacturer's function or the designer's function to educate the public in the conventional sense; and even less is it their function to shove ideas down the public's collective throat for which its collective mind is not ready. It is, however, one of the manufacturer's responsibilities, I would suggest, to respect the public and not to underestimate its intelligence or its ability to discriminate between good and bad. For, if all or most manufacturers and retailers catered for a fictitiously low public taste they would, between them, create a manufacturing monopoly for satisfying a taste worse than that which exists in reality and the public could do little about it; I should have thought that this was not a good thing because it is bound to frustrate the customer, harm the manufacturer and make it impossible to export the product to the advanced countries of the world where such monopolies do not operate.

A modern British woman's house, 1950–1970. The furniture is post-war 'utility'.
(*Design magazine*)

BOMBER

IVA BRITAIN AS I SEE IT

A great deal has been said in the past twenty-five years by people of all political persuasions about the need to rebuild Britain but little or nothing about the quality of the rebuilding. As a result, it is only a small exaggeration to say that, with few exceptions, all we have done is to replace old slums with new ones.

When we consider the beauty of our countryside and the fact that there is only so much of it, which, once despoiled, is irreplaceable, we have to ask ourselves if we have done as well as we should and could have done and if our plans for future building are as sound as they ought to be.

'Lower Broughton (Manchester). Engels was here in 1844 and said it was a slum. In 1974 it is still a slum.' From the *Observer*, 3 February 1974. (*Photographed by Cattani/Obs*)

The back view . . . (*from Saturday Night And Sunday Morning. Copyright British Lion*)

. . . and in Hackney . . . (R. *Barlin*)

. . . and in Glasgow. (*Ken Lambert*)

Industrial buildings in Bradford.

Industrial buildings in Lancashire. (*Ken Lambert*)

Waterloo Station: it could be any British industrial town at lunchtime.

North London.

Ebbw Vale in Monmouthshire. (*Peter S. Dole*)

Housing scene in Manchester (or Moscow?), *The Listener*, 30 May 1974.

The sad grey face of the North of England, Southern and Central Scotland. Countless towns in these areas look exactly the same and all of them look like the outskirts of Edinburgh where this picture was taken. Houses like these were built in vast numbers from 1945 to 1960.

High-rise flats near Edinburgh Airport, *c*. 1970.

Private-enterprise built houses in the northern outskirts of Edinburgh, 1973.

The old part of Galashiels built in 1860/70 with a more recent Tweed Mill in the foreground.

The setting of the new Langlee Estate, Galashiels, built in 1970.

Langlee Council Housing.

The plaque at the centre of Langlee Estate commemorates Sir Walter Scott and the view he loved. Below, the view today. The Eildon Hills are in the background, the Council Housing Estate, built in the 1960s, in the foreground.

The siting of Langlee Estate.

Small blocks of flats in close proximity in the Langlee Estate, Galashiels.

The playground, Langlee Estate, Galashiels.

Large blocks of flats in the Langlee Estate, Galashiels.

Terraced housing, front and back views, Langlee Estate, Galashiels.

V

However, we in this country may already have gone quite some way in this direction. Thus, I sometimes realise with concern that in a Britain becoming increasingly Marks-and-Spencerised, there are about eleven hundred Woolworth stores, four hundred Marks & Spencer stores, three hundred British Home Stores and about one hundred and fifty Littlewood stores; say, about two thousand stores altogether, situated strategically in all the important centres of population, selling identical things at similar prices (all of them, for all practical purposes, modelling themselves on Marks & Spencer, because it is the most financially successful of them all). The vast majority of us find it more convenient to buy our goods in these shops whether we think that they are excellent or merely acceptable. We have few alternatives for shopping elsewhere either in terms of time taken to find other sources of supply or in terms of price. Collectively therefore, and sometimes even individually, these four groups of people, still only a handful in number, the buyers of these stores, exercise a very real influence and a monopoly on taste in this country for we can buy only what they decide we should see on their counters. For practical reasons and in order to be financially safe, they, the buyers, have to reduce taste to its lowest common factor and cater for it and for nothing else. So, thinking as they do only of what they estimate the largest number of people want, and being able to achieve this safely only by not offering anything very different from what they have sold well in the past, we already seem to have a situation in Britain now where the more easily accessible shops all over the country, the most efficiently buying and therefore the most reasonably priced, are catering exclusively for the lowest common factor in taste. They look pretty much the same and sell very similar merchandise at similar prices. (If one were led blindfold into one of these stores and then, the blindfold removed, one was asked to identify the store, the task might well be beyond one.) The few people who unwittingly exercise this monopoly usually cannot reconcile what is financially safe with what is visually satisfying and sound.

An article in *The Times* of 8th January 1974 was a profile of two of Marks & Spencer's five managing directors and indirectly gave some insight into the company's aesthetic code.

'At Marks & Spencer,' the article said, 'the atmosphere has been described as that of a religious sect or ideological group. This is certainly the impression from talking to the company's two new managing directors. They talk of the principles established by the founder of the business; of a constant philosophy; the search for improvement; the consideration of human relations; the improvement of values.'

'A few years ago,' the article continued, 'a man wanting anything more adventurous than a white shirt, a pair of grey flannel trousers and a charcoal pullover was advised to look elsewhere. Today the range is greater but still providing for a clientèle sartorially conservative by contemporary standards.'

Or to put it differently, today the range is wider but anyone who enjoys his clothes, cares about what they look like (as well as what they cost and how they last in respect of which Marks & Spencer usually offer excellent value) is well advised to go elsewhere. Marks & Spencer assume that their customers know all about abrasion tests, about light and washing fastness and about the number of stitches per inch applied to a garment and that they deserve the best money can buy. They also seem to assume that these same knowledgeable customers are, unfortunately, much less interested in clothes which are visually appealing and exciting.

It would not be entirely unfair to say that, as a company, Marks & Spencer are interested in just about everything from artichokes to anoraks and when it comes to clothes they care about everything from planning to packaging; but they are not greatly concerned with what the clothes they sell look like. It is conceivable that by taking this attitude they do not serve their customers as well as they could.

In view of this, does it make sense that emphasis on tangibilities and on profits, linked to good public relations and staff and customer relations, should, regardless of the important intangibles and of the no less real aesthetic satisfaction their products should give, make this company highly acceptable, a paragon of virtues, praised and copied?

That today's Marks & Spencerisation of Britain is not an unmixed blessing (even though it is arguable that in its early days it was) becomes more obvious still when the effects of Marks & Spencer on their suppliers is considered. A supplier who depends for the bulk of his business on any of our chain stores automatically disqualifies his company for exports. Chain-store lines are not exportable, meeting, as they have to, the lowest common denominator in British taste. They are, compared to what is available in their own countries, relatively cheap for tourists who may purchase them when they visit Britain, for that reason. But export them in the usual way and add the various taxes and charges, and this advantage disappears and they become unsaleable. (According to *The Economist*, 26th May 1975, Marks & Spencer shops in Europe and Canada do not do well. After the first few weeks of opening – when the customers' curiosity has been satisfied – sales drop and never recover; this is quite a different trading pattern from that of Marks & Spencer shops in Britain where they do recover and where the turnover of each Marks & Spencer shop grows steadily.)

The effect on design and on designers of catering on a gigantic scale for an assumed lowest common factor in taste can be imagined. Think of all the designers in all the companies supplying two thousand retail outlets and ask yourself: are these designers designing what they have been trained to design or what they are told to design? Think of all the designers within the organisation of some of our chain stores (and some boast quite well-known names) and try to trace their influence on the design of the fabrics and the garments being offered. You will have little success. It seems to me fair to ask whether in textiles at least chain stores are the advance on the older methods of shopkeeping which we believe them to be.

Unfortunately, these issues are not raised or aired because it has been trendy for the past fifteen years or so to maintain that Marks & Spencer is the progressive place for shopping. Marks & Spencer labels are worn on sleeves. Really up-to-date books on etiquette advise the reader to make it plain at the first opportunity that though her hat may be from Harrod's, her pantie-hose is pure Marks & Spencer. No woman writer who wants to be considered 'with it', progressive or simply sympathetic to the toiling masses can afford to miss an opportunity of telling her readers that to be seen of a morning, in slacks, at Marks & Spencer is much more meaningful than airing the tiara at the theatre on a first night. One of them recently wrote an article on how to shop at Marks & Spencer: let everyone know that one shops at Marks & Spencer and yet, by spending a mere thirty-six hours per garment on altering those Marks & Spencer bargains, fox them all by looking as if one never went near there. (There is also another game, just as fascinating but played in the opposite direction: You buy your working-man's shirt and trousers at Yves St Laurent, opposite Harrods or in Bond Street, at about fifteen times the price that the working man would pay for it. You then pull out the threads at the cuffs and at the bottom of the trousers and fray the cloth at these points, have a bath or two in the outfit, roll on the pavement or the bottom of the garden before you take it off to dry and allow the sun to bleach it and only a few knowledgeable people will realise that you haven't come to the party straight from a day's work at the docks. An interesting reversal of rôles compared to fifty years ago; then the poor workers genuinely tried to copy the rich, now the idle rich phonily try to copy the poor.) It is all nice, clean fun but not very helpful in sorting out our thoughts on our genuine requirements, where to look for them and what to do about the general failure to meet them. Nor does it help to see, in the financial pages of our 'quality' papers, regular adulatory articles on the steady and growing financial successes of Marks & Spencer and about Marks & Spencer's own efforts to convert British industrial thinking to their views and methods. It is one thing to say that Marks & Spencer have been leaders in looking after their employees, in rationalising the relationship between the physically ascertainable qualities and the prices of their merchandise; with having led in service and in profitability. It is quite another to hold them up as paragons in the field of originality and design.

Just as we distinguish between the qualities and characteristics that make a tobacco baron in a prison, a merchant adventurer and a prince of industry, so we should perhaps differentiate between the qualifications needed to make profits alone and those needed to make or sell principled products and profits.

It has become quite common in recent years to see, in papers like *The Times*, articles discussing, on one page, the importance of good design from a personal and national point of view and then to see, on the opposite page, superlatives lavished on what is not just ordinary bad design but pretentious bad design.

A visit to the Blackpool Gift Fair (or, for that matter, to St Mark's Square in Venice) is an experience of undreamed nightmarish possibilities. The collection of rubbish, to which there are, of course, some exceptions, at this fair and in the Square is stupefying. The corruption of standards in the articles sold by the gift industry has to be seen to be believed. Yet, there is no doubt about it; millions of people actually pay good money for the stuff, millions of pounds worth of business is done every year at St Mark's, and at the Fair. A demand is being met, the wheels of industry and commerce are kept turning and it is all justified by the fact that the customer is getting what he wants.

Sketches from the Blackpool Gift Fair 1974. (*Glynis Edwards*)

In the light of these facts, it is not going too far to ask whether there are perhaps other criteria than what is usually meant by 'what people want' to determine the place and the function of design in industry, its relation to past trends, to current demand and to likely future needs.

One does not quarrel with the argument that it is the manufacturer's function to take the situation, in regard to demand, as he finds it and to satisfy it; with profit and, if possible, with pleasure to all concerned. The difficulty with this simplified approach to industry and to business is that in an effort to be safe and profitable the manufacturer will, in his assessment of current demand and future needs, trust his experience and impression of the past more than he will trust what his imagination and sense of adventure tells him will happen in the future. The sad, but predictable, outcome of an extreme case of this kind is a frustrated, desensitised, aesthetically debauched public and a surplus of ex-manufacturers.

VI

To underestimate completely the public's taste is individually damaging and stores up trouble economically for the future. To overestimate it completely is to get into immediate financial trouble. To do neither but to be just sufficiently ahead of public taste and demand so that the public can understand and appreciate one's aims and follow one and buy the product is, one likes to think, the ambition of all sensible people in industry.

Speaking as a designer and as a manufacturer, it has always seemed to me that many more people than one would suppose have a certain amount of natural good taste and were it not for outside circumstances, given a free choice, these people would quite naturally choose the well-designed object in preference to the badly-designed one. If I were asked to explain why I believe this to be so, I would, I am afraid, after quoting Tawney (page 27) launch into the following lengthy explanation:

1. It is accepted that children have an instinctive ability to see things with fresh, receptive eyes. Therefore they also have the ability to express themselves verbally and graphically in individual, sometimes original and beautiful ways. One needs only to look at children's drawings and paintings before they go to school and during their first years at school and to listen to them carefully, to realise this.

2. By the time the average child has spent two or three years at school and has had formal instruction in drawing and painting, that sense of freshness and individuality seems to depart from its work. The child has been taught to look at the world through adult eyes and the tinted and biased spectacles of convention have been interposed between the eyes of the child and the world at large. In the nature of things this is almost inevitable because education is partly discipline and must, to a large extent, make use of accumulated experience and knowledge and has to pour free, irregular

impulses into more or less conventional moulds. This is the tribute which convention extracts from progressive change.

3. Sound education and stubborn good sense, however, make certain that pupils learn how to use the spectacles of convention when necessary without losing the ability to use their own bare eyes.

4. Being resilient, human nature retains, through the subconscious, much spontaneity; which is another way of saying that, in spite of all educational efforts and social pressures, people often fall back on their instincts and rely on them for their reactions especially when they are free from pressures and relaxed or, at the other extreme, when they are wound up, when they are under tension and have no time to think.

Thus I believe, with a good deal of supporting evidence to help me, that most people react initially, when they are young, in a most uninhibited way to the visual world around them; and that even later in life they notice, everywhere, the larger, more obviously impressive things and sometimes the small ones too, of which the large ones are composed; that they are moved by what they see; and that according to their individual natures and abilities they do something about it.

The simplest, most direct and most primitive reaction is to want to possess an object which has provoked one's admiration and wonder. If this is not practical, for whatever reason, many of us try to take token possession of the object and will, in addition, or as a substitute, want to relate it to ourselves somehow; and most of us succeed in this whether we realise it or not. Regardless of education and background one of the ways in which many people find it easy to relate their desires to themselves is through their clothes and other objects in their personal environment. The beautiful and exciting things which they notice are first digested, abstracted and translated into more easily acquired things such as pictures, clothes and many other artefacts and these are then related to themselves by possession and wear or by possession and display.

Let us take, as an example, a family on holiday. They travel in their car and suddenly, around a bend, they come upon a beautiful view with blue skies and the sea; deep green firs reaching to the water's edge and separated from it by light, grey-beige shingle. Tucked away amongst the trees they can just see a white villa with a blue-green roof and the spot of colour here and there which the yellow and orange sunshades present. The sun is behind and slightly to the left of them and the view is in front and slightly to the right.

The first, probably unspoken and barely conscious reaction of the family is: 'This is beautiful; wouldn't it be nice to stay here.' The second: 'Wouldn't it be nice to live in a place and a house like this all the time.' (This is the wish to acquire.) The third: 'Let's take a nice photograph of it so that at least we can remember it and re-live this moment more vividly in the future.' The fourth (less conscious than the previous three) will be an attempt not to acquire the beautiful villa directly but to recreate this ideal somehow, somewhere, to a larger or smaller extent depending entirely on the means available and on the circumstances. The fifth, even less conscious reaction, but possibly the most important one, will be set in motion when the whole scene, the mental snapshot of the admired view, will have been absorbed into the subconscious and after a period of slow gestation and a considerable change will have become part of a repertoire of vague aims and ideals: in this instance the ideal is that of an 'atmosphere in depth' composed of textures, spatial relationships and above all of light and of colours. The same view will seem different to each individual who had seen it and it follows that the same view will be remembered differently by each of them, regardless of the 'physical realities' and of geographical and atmospheric facts. Whatever the impressions they retain and recreate, the deciding factor, the one which will occupy the largest portion of their mental memory picture, will usually be the light and the colours of the original view. In this sense most people are theoretical abstract painters or Impressionists and practical pipe dreamers.

We are all constantly and subconsciously on the look-out for instant new pleasures and for the recapture of past pleasures and good experiences. Most of us, when we are reminded of past pleasures, when our memory is triggered off by a smell, a tune, a view, a film, a painting, a woven or printed fabric, a carpet or a wallpaper and when we relate this pleasant memory to the object at hand, will unknowingly recognise the memory, associate it with and therefore like the object and want to have it.

We are, when we choose clothes, as on other occasions when we choose different things, often influenced by whether the clothes or other objects we happen to be choosing act as catalysts for pleasant associations or memories. Obviously, if a dress material reminds us of a beautiful view seen, of flowers or even of a whole garden enjoyed, or conjures up pleasant memories from the past in a general way, or causes us to associate it in some way with an admired person, we shall like it and want it. This is the fairly obvious corollary to my way of reasoning about our reactions to beautiful things seen in general. Simultaneously at work with this instinct for the permanent acquisition of a past pleasure or of an ideal is another factor. It seems to me that our ability to abstract impressions is not confined to the world around us but also applies to ourselves and to other people. Each of us has a very personal way of summing up other people as a whole, a special view of other people's personality and identity as expressed by their actions and appearance and we each have a very personal view of our own physical appearance in detail and as a whole. This impression is subjective and is the boiled-down sum total, the quintessence, of all our glimpses of others or of ourselves. We carry in our minds certain data of our abstract, impressionistic idea of ourselves and also of the people we know.

When we dress we are trying to be more than just practical; we are trying to choose clothes which will not only keep us covered, respectable and warm (or cool), but clothes which will somehow help to make others see us as we should like them to see us. We choose clothes which will make us look better, feel more self-confident and which will give us and those looking at us an indefinite, pleasurable feeling that cannot be measured, or explained in simple terms; but which will, in combination with ourselves, as we see ourselves, either give us something which we do not possess at all or something which will enhance, bring out and emphasise that which we have in certain quantities but which we judge not to be sufficiently clearly in evidence.

It seems to me, therefore, that one of the challenges which we try to meet and the pleasures which we obtain from choosing clothes is the act of juxtaposition of two abstract, impressionist compositions, physical images or colour concepts: one, our own image of ourselves, and one the abstract image of what we may have seen at some earlier point in time. We try to combine these two compositions in such a way that they activate each other and that the sum total of both is visually more meaningful than the straightforward added value of the components would be.

Now I am quite aware that people in general do not attend courses on 'The Abstraction Of Views Into Memories' or on 'How To Be A Mental Impressionist' and certainly not on 'How To Marry Your Image Of Yourself To A Pleasant Memory From The Past'. Nevertheless I think that this process of seeing, sensing, rearranging and superimposing goes on all the time in most of us. We all have our own personal impressions (compositions). We all combine them in many ways and success is the tolerable fit of two originally unrelated, abstract compositions. The same processes can be said to be at work when we buy a house, furnish a room or choose a car that will mirror ourselves and convey to others something about ourselves which we want to say but cannot say in any other way.

This then is some of the background thinking to my personal experience that most people, particularly women, will, when faced with an aesthetic choice, choose the good and reject the bad. Anyone who has watched women choosing cloth, clothes, hats or cosmetics – all things which help the presentation of oneself to the world – will have seen that they enjoy themselves, that they concentrate on the task in hand to a point where all other cares seem to be forgotten and that, while they seem to be in a trance, they are actually using all their visual senses, and some other senses as well, to their limits. It can also be noticed how most of the time their eyes are focused on the middle distance and they seem to be relating the objects (cloth, clothes, hats, jewelry) which they choose not so much to themselves, there in front of the mirror, but to someone who is not quite there and exists only in their imaginations.

It is my belief that most people will instinctively go for the aesthetically sound, for the well designed, the well proportioned and above all the well coloured, if they have a choice. There are often real, practical reasons why something which is desirable in a product cannot be incorporated into it, and usually it is price. But I am fairly certain that if good and bad design are offered at the same price or even within ten to fifteen percent of the same price, people will buy the good in preference to the bad. Hence my 'Bee-line Theory' of the thousand dresses; and my recognition of the validity of the argument that it is not industry's function to educate public taste. There is no need to; public taste is ahead of the taste allowed it by manufacturers and reflected in current manufacturers' products. Even in our present circumstances, quite possibly not perfect, given the choice, people would choose many more good things, well designed things than they actually do choose because now they can choose, broadly speaking, only from amongst badly designed things.

The point can be made that it seems contradictory to say one minute that women will instinctively choose well – the colours and designs which suit them – and then immediately follow this by saying that it is wrong, when creative activity is involved, to anticipate demand by polling potential customers and then producing what the majority seems to want. If people know what is 'good' why not let them tell designers and manufacturers about it so that we can make it for them? In practical terms this attitude could be boiled down to a manufacturer's credo of: 'Find out what *they* want. Give them what *they* want. Get the hell out with what *you* want.'

This is, I know, putting it crudely; but then the question stems from a rather crude way of looking at one's work. If we agree from the outset that there is more in doing a job than merely making a living, we can start answering this point by comparing the function of any talented person allowed to develop his talent by training and earning a living by selling this acquired skill to the public, to that of members of a government elected to do a job, (for which one must assume they have both talent and experience) and so allowing them to 'practise politics'. Just as a government must carry out its own programme and cannot constantly run to and fro between public opinion polls and Whitehall if it is to function, so a designer, once given a job, cannot depend for his decisions on what the public thinks. Both the politician and the designer, once they reach their positions, must find their own solutions to their respective problems, using their special qualifications in the process. Qualifications, by the way, the public do not have. The penalty for both politician and designer, if they get out of touch with their grass roots, is the loss of their jobs. We accept Parliament's rôle to lead rather than to follow public opinion. (Recent instances are legislation on capital punishment, on immigration and on the Common Market.) This, in spite of all the clamour and the well orchestrated noise made by leaders of sections of the public who oppose innovation and the things they have not experienced before. In public companies we accept that it is the board's responsibility to manage without polling the shareholders on the day-to-day running of the company or indeed on anything except sometimes in respect of fundamental changes in company policy or constitution. Is it then too much to ask that designers should, in matters of design – where the public knows only what it likes but cannot create what it likes – be allowed and encouraged by manufacturers to lead without the necessity of repeating with very minor variations what everybody knows has sold and has been found acceptable in the past?

It is amusing – when it is not infuriating – to have to argue for the very real concept of the Freedom to Choose (the components which combine to make up the mainly physical whole of everyday living) with people who are prepared literally to lay down their lives for the abstract concept of Freedom (in regard to the ideological components of everyday living). It is less amusing, because it is sad, to have to argue for the dignity that comes only to those who respect their own physical environment (which starts just outside one's skin) and their own work. No constitution, as far as I am aware, guarantees the right to dignity; yet dignity is an important ingredient of the total concept of freedom, and personal dignity depends to a large extent on the context in which we see ourselves and present ourselves to others; in other words on clothes, on homes and on places of work and leisure. In a sense Art, in its struggle for the heart of matters, rejects all these props to dignity and relies for it on the harmonies of the naked body. Nevertheless the props are real and are needed by most people most of the time and one would have thought that there is little room for argument either about their basic importance or about the need that they be available in an individuality-enriching and encouraging variety.

Few of us in Britain feel happy about quite a number of political systems throughout the world because we are convinced that were the people in certain countries free to choose they would choose differently and better. What matters to us is that people should clearly be confronted with alternatives and be able to choose freely. We believe that once the principle of Freedom is maintained everything else which matters to a nation and to the individuals who make up the nation will follow sooner or later. It is no exaggeration to say that in normal times the freedom to work and the freedom to spend one's earnings (after tax, of course) in any way that one sees fit are important parts of our idea of Freedom as a whole. Thus, in spite of our acceptance of certain national monopolies, a chain of government-run shops with a monopoly on any one product or on any number of products is quite inconceivable to us. It would be unacceptable even if the government offered to hold a poll once a month to find out what the people wanted and undertook faithfully to supply it. We know that it would not work. We know, even if we have not quite said so, that we get better results if the people who have the talent and training in art, science, technology and commerce are allowed to combine in industry, are given their head and left free to offer the public a choice, through competition. We think of our system as clearly less dangerous to freedom than a system which is described as the expression of the 'people's will' and we prefer it to the imposition of restraints on the creative section of the community.

The difference between the school of Social Realism in the Arts and Crafts as it is enforced and is practised in, say, the Soviet Union, and the various trends in Art which freely and frequently come and go in Western countries, is almost too obvious to be mentioned in this context. In Russia, the artist's creativity is quite deliberately suppressed and allowed to function only in a way

(*Sunday Telegraph*)

the Government considers desirable and in the public interest; the artist can create only that which 'the public wants' or what the Government thinks the public should want and therefore only that which the public can immediately follow and understand on instructions from its rulers. In the West, artists are free, indeed expected, to create what they want and are then often helped to persuade the public into wanting what they have created. The contrasting effects of the two systems on Art and on Industry can be seen plainly and do not need to be discussed here.

It puzzles me therefore that we in Britain should take ideological and political 'good sense and good taste' in ourselves and in everybody else for granted and assume that we all want freedom and the other democratic goodies, but that we should, in spite of all the evidence to the contrary, assert that aesthetic bad taste is the chronic condition of humanity and that in order to be viable industry must cater for it. Surely human beings exercised aesthetic discrimination – especially in clothes – several millennia before they started choosing the political systems by which they wanted to live. Is it not more likely that our aesthetic judgment is better developed and therefore to be trusted more than our political one?

VII

The real standard of taste of the British consumer is, I think, higher than it is assumed to be by the manufacturing and the distributive industries. The British consumer can, however, choose only from among the things that these industries put on the counter. It is, therefore, mistaken to say that here in Britain demand determines supply. The position and the nature of British industry and commerce is such that exactly the opposite is true; it is supply and the people who control it, who determine demand (albeit not entirely to their own commercial satisfaction); and if these people, the suppliers, lose touch with the rest of the world, as it is possible to argue that British industry seems to have done imperceptibly but surely over a period of time, they are still able to sell their products here at home because they are in one way or another protected; but they cannot sell them to the rest of the world.

The hidden, real standard of demand in Britain which is not being catered for, is higher than the assumed standard. Comparisons show however that our real but hidden standard of demand is still lower than the real standard in other, advanced countries. This has obvious disadvantages in personal and in national terms and it would therefore be worth, in the long run, investing in the improvement of our real standards. This change could be brought about mainly by a new approach to education at all levels, for an improvement in education will automatically mean that the consumer and the producer of the future will be better educated to fulfil their respective rôles. Admittedly, this is not initially an industrial function but a political one. However, the persuasion of politicians to bring about changes for the better, to the benefit of individuals and of industry, is everybody's function, including industry's.

Having earlier dismissed aesthetic education from the immediate sphere of industrial responsibilities, I should now like to return to it and look at it briefly in the context where it properly belongs:

as a part of education in general and of education in visual matters in particular. It is this aspect of education that affects the quality of life and the economic well-being of a nation to a far larger extent than seems to be realised.

It has been mentioned that other countries, those where the standard of demand as regards quality is higher than in ours, have not bumbled into their present, happy states by accident. In addition to certain favourable circumstances, they have made consistent and carefully thought-out efforts over something like fifty years to instil into their populations knowledge and discrimination in matters of design and in enjoyable, full, sensitive living. The obvious conclusion to draw from this would seem to be that if we want – be it only for purely business reasons – to improve the pattern and the quality of demand in Britain so that our mainly home-consumed products become more exportable, we shall have to think of and implement some similar programme sooner rather than later. One would also be right in assuming that even if such a programme were worked out and introduced tomorrow, the results would not be felt for at least ten years, perhaps fifteen or even for twenty years.

This being so, it is also fair to conclude that it is in industry's interest to help those responsible for education in general who are, as a rule, not too familiar with industrial and economic affairs, to draw up plans which will help healthy industrial development, as distinct from unhealthy industrial development, since development per se is not necessarily desirable. As these industrial aims are not only not in conflict with the broader aims of education but are actually part of them and help them, industry needs hesitate even less in impressing their urgency on the educational authorities.

It seems to me to be vital and urgent that we change our thinking on the way we teach art in our elementary and secondary schools. I imagine that eventually we may have to introduce one or more new subjects on Seeing, on Using and on Living. We may have to change our ideas about art schools, about the kind of art teacher we train here; and our ideas about the training of artists and of designers. As to the Design Council, I would suggest that its aims should be reconsidered, its constitution reshaped and brought up to date and that its methods of operation be changed; we should try and retain the good, eliminate the useless and the bad and introduce a number of better ways of doing things.

We should, I think, also note the lack of realisation by people, and therefore by all political parties in Britain, of the close relationship between education in visual matters and the viability of Britain as an exporting country to the developed and the progressive part of the world (not to mention the validity of Britain as a country of genuine, human values). All parties promise, if elected, swift if not instant salvation. This particularly applies to the economic concerns of the country. 'Elect us,' they all say, 'and in one, three or at the most five years all the country's financial troubles will be over.' No one who realises the real roots of our troubles – the attempted aim of the first two chapters of this work – can possibly believe that there is a short-term solution to our problems. All who sincerely promise and indeed sincerely try to find a short-term solution are bound to fail, if my conclusions are substantially right. And all those who, purely by social manipulation, try to change the nature of our industrial output are quite simply ignorant of industrial realities and of human nature. The quality, quantity and the content of our industrial output can and should be changed but this should be done by those who have talent and training for it and not by politicians and social agitators who have neither.

There is nothing in the taste of the British public and in the ability of British industry that a well-thought-out educational programme on all levels cannot put right in, say, fifteen years. When all is said and done, all the public should have to do is to choose between a number of acceptable alternatives; and the ability to choose well, though it is similar to creating, is nevertheless infinitely easier and therefore an attainable objective.

If it is right to think of industrial creativity as not wholly unlike artistic creativity and if it is not wrong to classify industrial design as essentially the same in kind as design in the abstract, then the importance of putting first things first begins to be apparent.

The basis of any worthwhile industrial cycle is the Idea; by definition new, different, as yet unwanted because unknown. If the idea is sound then production can follow; and sales will be based on the public's ability to choose well, given a set of alternatives. This is not the same as saying that the public can design the product that it wants or even that it should help to design it. There is an important difference between knowing what one likes when one sees it and being able to imagine, in practical detail, something different, desirable, never having seen it. The first is intelligent consumption and the qualifications for it can be acquired. The second is originality and creativity and the qualifications for it are gifts which can be developed but cannot be acquired.

It seems reasonable to say that one of the key factors to intelligent consumption, to worthwhile and significant industrial production, to personal satisfaction and to a healthy national economy, is the mentality of manufacturers and the public. There are a number of forces which will affect and change any given climate of opinion. Education is probably the most potent and the longest lasting of them, even if it may be the slowest.

It has been my intention here to examine and to establish the reasons behind the view that in industry, as in art and in personal matters, having a standpoint is preferable to not having one; that it is right to defend it, sometimes even to propagate it without blindly adhering to it; and that the alternative of making a point only of not having a point of view, except that of being all things to all men, can rarely be justified.

In designing as in certain other fields of activity where there are no absolute standards against which to measure the validity of the work, ambiguity and compromise will often, perhaps almost always, weaken and in the end destroy, what they seek to save and to strengthen. An uncompromising determination to succeed is not enough especially if one compromises in everything else. On the face of it no two professions could be more different than designing and politics, yet ultimately both the politician and the designer are persuaders (both try to make people see the world through their own eyes) and persuaders in particular have to be careful not to compromise what they stand for in their anxiety for conventional and quick success.

This is why I think that an experience narrated in 1973 by Christopher Serpell, the BBC Correspondent, may serve as an apt summing up for this chapter. He said:

'Fourteen years ago, I was included in the press caravan which travelled through the Soviet Union alongside Mr Richard Nixon, then the American Vice-President. Before we left Washington, Mr and Mrs Nixon threw a "getting-to-know-you" party at their home for all us fellow-travellers and although I was still waiting uneasily for my Soviet visa – delayed as usual till the last possible moment – I went along for my first personal contact with the Vice-President. Apart from a welcoming handshake, that encounter didn't really happen until I was leaving the party. "Chris," said Mr Nixon, intercepting me as I eased my way towards the front door, "I've something particular I want to say to you." Slightly mystified, I was drawn into seclusion behind a large potted plant. "I've got an exclusive news story for you," he went on, "which I believe might be of interest to your British listeners." I made the appropriate gratified noises. "As you know," said the Vice-President, "during my tour of the Soviet Union I shall be making a number of important speeches. I felt the British public might be interested to know I shall be modelling them on the style of your Winston Churchill." He paused impressively. This was my scoop.

'It was never broadcast, of course. Somehow, I couldn't see myself convincing the British public, in the year 1959, of the significance of this curious piece of information: and indeed when Mr Nixon made his speeches in the Soviet Union, they bore no discernible similarity either to the oratory or the attitudes of Sir Winston Churchill. At first, I thought Mr Nixon had singled me out for this peculiar confidence out of the kindness of his heart: seeing me, the odd Britisher out in an all-American gathering, he felt I should be equipped with a British-style news story. Later, I realised that this was just one more move in his indefatigable campaign to cover all the angles. If there was a British reporter there, interested in the Soviet tour, a British angle on this event must be provided.

'It is, I think,' concluded Serpell, 'this desperate attempt to cover all the angles which has now betrayed Mr Nixon.'

VIII

To have a point of view, to be prepared to live by it, to defend it or even perhaps to propagate it, is not, in itself, of over-riding importance. Many thoroughly bad and even evil points of view have been and are being defended and bravely fought for; that does not change them into good views. What matters is to hold, stand by and review informed opinions based on up-to-date facts, arrived at after open, broad-minded consideration.

It is therefore important to be educated, cultured, civilised, sensitive in regard to all the senses, responsive and alive. These pennants of humaneness should be flown high for their own sakes. If they are, it will be impossible for them not to be reflected in every aspect, including the industrial aspect, of living. It is clear and beyond doubt that education is one of the key factors in the shaping of individuals and of nations and it is a controllable factor at that. It may therefore be worthwhile to look in some detail at the rôle it plays and could play in our industrial future.

Being educated in Britain today implies being literate and being numerate; a 'good education' implies being cultured and civilised in addition to being literate and numerate. Literacy is the ability to read and write but it unavoidably includes some awareness of classic and modern literature; it demands the ability to speak in a clear, accurate and, if possible, a graceful manner. To be numerate one has to be able to understand and to deal with numbers and with the simpler problems which involve numbers. Literacy and numeracy are the minimum educational requirements of our society because without them communication of the most simple kind for today's requirements is impossible. General education is therefore primarily concerned with the training and with the development of the mind for utilitarian reasons and for practical purposes. Since the mind functions through and with the aid of the senses, it is impossible to train the mind and to leave the senses completely untouched; to this extent the senses too are involved in most of the educational processes. The training of the senses and their development, in the context of ordinary education, is however completely dependent on and subject to the requirements of literacy and numeracy.

A good education usually goes beyond the strictly utilitarian needs of earning a living and aims at producing a fully rounded person whose brain will be trained for the performance of certain necessary, useful tasks but will also contain seemingly useless knowledge (which nevertheless forms part of a complicated and effective piece of equipment for dealing with the problems of living, if not necessarily with earning a living) and one or more of whose senses will also have been trained to perform their functions better than if they had been left untrained. A good education aims, therefore, at the strengthening and refinement of as many as possible of the senses in the individual, the development and growth of the intellect and at their coordination and control in the context of an advanced level of social development.

The strengthening and the refinement of the senses for their own sakes are comparative luxuries, not indispensable for economic survival. Their growth is therefore considered to be not quite as important as the development of the mind in the interest of literacy and numeracy, and only well-off societies can allow themselves the luxury of including the training of the senses in their educational programmes.

Education in itself, therefore, implies little in regard to the development of the senses; culture and civilisation do. They imply that the basic necessities for living have been met and dealt with on an adequate level for the day and that enough energies and resources are left over for the comparative luxuries of an apparently non-essential kind, for the bringing into an advanced state of development of one or more of the senses with which we are endowed by nature (sight, hearing, taste, smell and touch).

Of the five senses the sense of sight is probably the most precious: vision is man's primary sense. A blind person will find it more difficult to acquire the educational weapons of literacy and numeracy, necessary for independence, than someone deprived of any one of their other senses. The sense of sight is an 'over-riding' sense because even when it does not dominate an activity it forms the background without which the other senses find it difficult to function. If civilisation implies the refinement of some or of all the senses, the sense of sight must be at the top of the list, the most obvious candidate for refinement; and the process of civilisation, when it starts, should logically start with the sense of sight.

It is not, however, in the British tradition to make much of or to worry unduly about visual literacy, about developing and civilising the sense of sight especially in regard to objects which relate to the person. It is possible that the self-deprecatory nature of the British character finds it embarrassing to bother about one's appearance in fussy detail. Physical appearances seem to matter more in many other countries. Not that the British attitude is without its charm. In gardens, for instance, the waywardness of the herbaceous border in contrast to the rigidity of geometric French arrangements, in houses, the informal, lived-in look of the British middle-class living-rooms in contrast to the tidy and austere Scandinavian or Central European look; and in clothes, the understated but carefully contrived air of a happening of the well dressed British male in contrast to the measured compositions of the flamboyant Italians, the over-stated opulence to which the Swiss and the Germans are prone and the sometimes painfully well scrubbed look, the tidiness of the Scandinavians.

However, once the tips of these submerged national characteristics and the hidden charms of British attitudes are acknowledged, the lack of an interest in visual literacy and in the visual aspects of current civilisation as a whole in Britain become obvious – and the harmful effects of this attitude on a purely practical, industrial level, on manufacturing and trade also become clear.

We are quite used to seeing actors on television as well as sports commentators, pseudo-personalities and detached professors, trendily, carelessly or clownishly dressed (the 'TV tie' will, I am sure, be an important point of reference for future anthropologists studying our era); this is the general British experience and not unduly worrying. Television announcers, who often receive plaudits and awards for their clothes-sense and who appear on the screen in jacket, shirt and tie (fortunately trousers are invisible) – doubtlessly hallowed family heirlooms enlivened by the latest from Carnaby Street – we can easily take in our stride. But to see responsible politicians and important industrialists (one does not obviously expect much from irresponsible politicians and unimportant industrialists) appear in the Seventies dressed in what seem to be, cannot be anything but, post-war utility suits with wide tram-line stripes, or in the checks or chalk stripes of the Twenties, with wide lapels and even wider shoulders, worn over tartan or multicoloured striped shirts and with ties splattered with polka dots, which are interspersed amongst thunder and lightning on a background of stormy seas, is too much and was only just bearable thanks to the tenderness of the black and white screen.

Such regular shocks to the eyes make one wonder how a nation which spends more on private education than any other can be happy with educating but not visually civilising its young; how a country which depends for its livelihood on selling its industrial products to the rich and to the poor all over the world (and in comparison to Britain increasingly to the rich rather than to the poor) can be content to bring up its young to be literate, numerate and ambitious without equipping them with the means to satisfy their customers, who are not only literate, numerate and ambitious but also visually civilised and to whom the visual standards of those whose senses are underdeveloped are unacceptable.

The relative lack of visual literacy in people who have had a small fortune spent on their education seems in itself totally inexplicable. When these same visual illiterates spend thousands of millions annually and have the final word on what our factories and products will look like, one starts to be seriously concerned, not only because of what these decisions mean in human terms to the individuals who have to live and work in them but also because one can see that the emphasis is on machinery and not on what the machine makes (a building in this context is both a machine for other machines or for people to function in and also a product of other machines which impinges on its surroundings). It is perhaps not sufficiently clear to those in charge of industry and of education, and to politicians (sadly, one has only to look at so many of them and to assess their appearance to see that it is not) that when it comes to consumer goods one cannot sell visually unappealing or even ugly things to customers whose senses are further advanced than the producer's; that one cannot sell visually primitive industrial products to visually civilised, sophisticated, industrial nations. Industrially, as far as the look of things is concerned, we seem to be unprepared to cater for other people's standards or to evolve our own to a level where others will be prepared to follow us. One suspects that this handicap is imposed on us from the start by ourselves, during the various stages of our education from primary to secondary schools and at the higher education levels; if this is right then there is something we can do about it.

The present system of education in Britain seems to be built on a number of principles. First, that education is a basic necessity like food, clothes, health and shelter to which everyone, regardless of effort, is entitled in moderate quantities and in modest but life-supporting qualities. In this we differ little from other countries.

The second principle is to look at education pragmatically and as an industry and to regard it as we regard the food or the clothing industries and soon now the insurance industry; most of us see no reason to stop people from privately buying, if they wish to, a different kind of general education from the standardised one provided by the state (and, as in food, drink and in clothing, in education too, the label often matters more than the product). In this we differ from most other countries who do not have our public school system.

The third is not to have any 'public' universities or art schools. So, when it comes to higher education, we are no different from other countries: students qualify for entry, broadly, by merit but also by other, less clear-cut or indeed democratic means. Our declared and accepted principle in regard to higher education is that it should be available, more or less freely, to everybody who can benefit from it.

The fourth principle, not directly relevant to our argument here, used to be to underpay all teachers, particularly secondary-school teachers, in relation to other workers in the community; in this we differ from other civilised countries.

When it comes to art and crafts (in a broad sense) in our elementary schools we generally adopt the principle that it is so useful a subject that we can and should devote more time to it than to almost any other single subject (a complete change from the way we looked at things, say, fifty years ago). We accept that art and crafts can help children to explore their environment, to express their views more easily and better than they can verbally at that age; that it enables them to communicate better, albeit at a shallow level, at a stage in their lives when their verbal tools are unequal to the task; that it is one of the best means in the training of the co-ordination of the imagination, the intellect (the mind) and the senses (the eyes and the hands) – and in the acquisition of some skills; and last but not least that it is one of the easiest and best ways of introducing children to both self-discipline and class discipline.

The name of the subject – art and crafts – is however misleading and causes a great deal of waste and confusion both to teachers and pupils. Much effort is nominally spent in 'making art' and in 'creating' it instead of in exploring and finding out, which would be much more to the point. For neither the teachers nor the pupils are, in the strict sense, concerned with art and with creation. Primary school teachers do not receive special training in art and therefore cannot teach it. The children themselves are primarily involved in obtaining pleasure by satisfying some simple urges such as the exploration of raw materials and are not directly concerned with the thought, discipline and the control which are the bases of any artistic activity.

It would be much more useful not to call the work that is done at primary schools art, but something else. We could, for instance, usefully call such classes 'environmental familiarisation' and include a course of a critique of artefacts and call them 'general awareness programmes' or 'visual-value programmes'. The fact that a person draws or paints does not automatically mean, far less guarantee, that he or she is an artist or that he or she is, in broad terms, visually aware. Drawing and painting in themselves have little to do with art; and artists are not necessarily or often visually aware outside their narrow spheres of activity. A look at most living artists, personally and at their immediate, self-created environments should prove this point.

At secondary-school level the relative importance of art and crafts to the other subjects taught gradually changes. In the first two years, that is while the child is eleven to thirteen years old, art and crafts is compulsory and an average of three hours per week are spent on it at various degrees of intensity and effectiveness, depending on the school and on the teachers. But then the streaming starts and the first signs of 'Life is Serious, Life is Earnest', with the implication at least that art classes are frivolous, irrelevant and fact-less, begin to show.

Great play is made in the official reports (see Appendix C) with the rôle of a 'rich art education' and with its effects on the sensibilities of children. But the number of hours which most children from thirteen onwards spend on art is not enough to have any effect on their sensibilities; and the principles involved in the teaching of the subject and its presentation are open to question. In other words, even if three hours per week were enough to inculcate a sense of awareness, sensibility and judgment, 'art and crafts', as taught at our schools, is not going to realise the aims of a 'rich art education'. Art and crafts classes from the age of thirteen onwards are restricted to those few who intend to make art their career one way or another; to those, more numerous, who are simply interested in it and good at it and enjoy it; and to those, even more numerous, who are deemed to be too unintellectual to achieve pass marks in the more academic subjects but who, nevertheless, have to be kept at school until they reach the legally recognised wage-earning age of sixteen.

In general, art and crafts is not looked at as a positive, constructive subject but as a 'holding' one. If there are unexplained gaps to be filled, perplexing difficulties to be shelved or time to be passed, art and crafts is brought in as a general, all-purpose palliative; and 'creative' art and crafts at that, not informative and factual, aimed at developing sensibilities and discrimination, but one which is to all appearances concerned with creativity.

Less than forty per cent of pupils continue to attend art and crafts classes in our secondary schools. The majority do not take any classes in art from the age of thirteen onwards and never formally return to look at or to think about art and crafts for the rest of their lives. This in itself is a startling fact because it makes us see the sixty million Britons with fresh eyes; most of us, it is clear, have had no contact with art and craft even tangentially and in the conventional sense for a very long time indeed and only a very limited and limiting contact with it until then. What standard of visual sensibility is it then reasonable to expect from most of us?

What of the minority, though – a fairly substantial one – which continues to have contact with art and crafts after the age of thirteen? Can one expect better taste and discrimination and a materially higher degree of sensitivity from them? Alas, the quality and the concept of our so-called art and crafts classes make it rather unlikely that, barring accidents, a fully rounded individual will emerge from our schools with senses responding to the daily challenges and stimuli in a satisfying, enjoyable, mature and cultured way. The reason for this is that we have not defined clearly enough in our own minds the meanings of creativity and of sensitivity – the difference between creating and re-creating – and, as a result, our system of teaching art and crafts is confused and not as effective in practical terms as it could be.

IX

To explain the difference between creativity and sensitivity we have to ask ourselves some fundamental questions. I know that when I choose a painting, a tea-set or a tie, I am bringing into play certain faculties which are partly natural and spontaneous and partly influenced by circumstances and by training; I am deploying all my feelings, my stock of information and my faculties of judgment in regard to colour, shape, space and texture in order to choose from a number of alternatives. I then proceed to use the item of my choice in combination with certain other items to achieve a new composition of objects which, when seen together, form a new entity (the painting in a room, the tea-set on a tablecloth along with other objects and the tie with a suit and a shirt). This exercise of selection and of the combination of a number of selected items is certainly the gearing of my sensitivity to the visual world around me and could, with a little stretching of the meaning of the word, be called creativity, which, up to a point, it probably is. It would however be more accurate to describe this act of choosing and the act of the subsequent combination of items as mainly a response to other people's creativity: the use of my sensitivity to receive the message sent by the creator of the object of my choice. To receive this message I have to have a receiver of sorts; without it, the message would be lost. My sensitivity is such a receiver. To react to what others create is to receive. To repeat – albeit with variations – the message which I have received is to identify myself with the contents of the message. To send fresh messages, on the other hand, is to initiate: to create.

When one asks, therefore, exactly what it is that makes one paint and sometimes design, one realises that certain things must excite people when they see them; and that they then feel that they have to do something about this excitement inside them; and that the only thing they can do is to paint the view, or the flowers, or the object – whatever it was that caused this particular feeling of excitement.

To paint and to design are, to certain people, the only ways they know which will relax them and bring them peace after this particular kind of visually caused tension.

Certain flowers may have this effect on individuals more than others; let us take tulips. When one sees a bowl of parrot tulips and gets excited about them, one may see them first, I suppose, more or less as anyone else would see them; but the moment one starts painting them, something happens and one begins to simplify the shape of each flower, to relate the shapes and the positions of a number of flowers to each other, to relate the long, thin stalks to one another and to the vase; one imagines and, if possible, arranges it all against a special, pleasing, related or contrasting background and all this happens without much conscious thought – almost automatically. In addition, probably a number of other things also happen while the painting proceeds; a texture may suggest itself as a result of the strokes of the brush or the palette knife; this texture is identified, fixed and elaborated on; developed and made into one of the rhythms of the painting; a new colour appears as a result of many mixings and juxtapositions – themselves the result of a certain amount of consideration – and one uses this colour, which in some way is quite new, as the underlying theme for the whole painting, to make parts of the painting relate and integrate. By now the whole concept of what one is painting has probably changed if compared to the impression that set one off when one first saw the tulips; what one is painting now are no longer groups of, say, pink, veined petals, filmy textures and precise, delicate shapes pushing upwards, sideways amid green, spearlike spikes; nor swaying cups with serrated, gimped rims. Somehow, between the moment one first saw them and the moment one applied the first dab of paint to the canvas and increasingly as one went on, the actual tulips have changed into something different; they have become an exciting, new arrangement of colours, shapes and textures which, on the face of it, bear only a slight resemblance to the tulips one might appear to be painting. In fact one is no longer painting those tulips as one first caught a glimpse of them, but certain aspects of all the tulips that one has never seen. In addition, one is no longer painting just tulips but certain other ideas which the tulips have caused to come to the surface of the mind – especially certain colours and textures which they suggest but which they do not actually contain and to which they do not strictly correspond. In fact, objectively speaking, one is not painting tulips at all; one is capturing a state of mind – brought about by seeing some tulips. One is sending out a tulip-based message; the starting point is the immediate personal reaction to something seen and then new discoveries follow; what one is recording are reactions and sequential thoughts engendered by a group of colours and textures and perhaps shapes.

If we think of a designer we might find that many paintings and designs are based on parts (sections) of paintings first arrived at as I have just described. Some artists see paintings in two ways; they have to add up to an interesting, coherent whole; but sections of them – anything from two inches square upwards, depending on the painting and on the implements used – must also make sense, on their own. So they may isolate such sections in paintings and enlarge them, first mentally, then physically. A designer or painter will work on these ideas and extend them so that he can reach new colour combinations, new textures and new shapes.

What I have said can be boiled down to this: that when I think about it I find that the work of a designer or of a painter can be divided into phases. First there is confrontation with objects. The stimulation and the excitement. This is a perfectly normal and frequent enough occurrence as far as most people are concerned and a strictly passive, non-creative, if enjoyable one; one is receiving. Then there may be the wish to put it down in a perfectly conventional, accepted way – more or less accurately, more or less realistically – as a token of a pleasure enjoyed or as a means of exercising skill or of carrying forward a pleasant experience into the future. One is identifying oneself with things seen and showing that the message has truly sunk in. Then there is the compression and the harnessing of the excitement into further constructive thought; thought into visible abstraction; say into painting which deviates from conventions. This brings about the surfacing of a number of ideas which have probably been lying around unnoticed and unused at the back of one's mind. These ideas are combined to varying extents with the object that has caused the excitement and this is where, I think, that receiving, mirroring and copying stops and creative activity starts. There may also be, later, the complete abstraction of certain parts of the painting. These abstractions can sometimes be used as the basis for designs.

Klee has said that the object of art is not to reflect the visible but to make the invisible visible; what I suppose he meant was that to be creative one has to have something of one's own to say; to be original, not merely to copy, no matter how accurately, what nature or what others have created; it means that excitement and its satisfaction are generated not merely or mainly by 'receiving', absorbing and mirroring the seen object but by the making and by the 'sending out' of new concepts.

This, I think, is in what creative people in a number of disciplines are involved.

I have used the word 'creative' several times and it is important, I think, to explain just what I mean by it. 'Creative' has become one of the 'in' words of our times and has, in the process, lost most of its meaning because it has been inflated and so, inevitably, devalued. Everything from planting cabbages to catching butterflies, from shouting slogans to smoking pot is somewhere or other, sooner or later, described as creative. In itself this would not matter much – it happens to many words – were it not that it is a key word in educational thinking and writing and it is also a key word in much therapeutic work and literature. Obviously it is important to agree on the exact meaning of this word in relation to activities in the description of which it is employed; in our case it is important to know exactly what we mean when we say that someone is being creative when drawing, painting, sculpting or when doing anything by hand, related to these activities.

It seems obvious to me that two people may be painting the same landscape, standing side by side, using identical canvas and paint, yet one of them may be creative and the other may be exercising to the full his techniques of merely observing, noting and recording. One of them may be producing a masterpiece and the other a chocolate-box top. (The implication that the masterpiece will end up in a museum and the chocolate-box top in a rubbish bin can, of course, be questioned nowadays when anything from worn tyres to minutely accurate paintings of tomato-soup tins end up in museums; but let us consider all the worn tyres and tomato-soup tins which do not end up in museums and allow for the possibility that a masterpiece will not be confused with a worn tyre or with a chocolate-box top and that indeed it is more valuable than either.) Both these painters may well have been trained at the same art school, yet nothing will make the one creative and nothing will induce the other to paint chocolate-box tops. Even more im-

A time of changing values for art and foodstuffs. (Marc in *The Times*, Friday 6 December 1974)

portant, both may well, for themselves, derive the same amount of pleasure from their work and transmit the same amount of pleasure to others through their work. We have to ask ourselves, nevertheless, should these two painters have received the same training or could each of them have been better and more efficiently trained if their differing talents had been recognised at an early stage? Could a creative person realise his potential better if he was trained to create and the non-creative person if he was trained with the talents that he does possess, in mind? Also, does the average person understand the differences between the work of these two hypothetical painters? Are they both creative?

People (reports and committees in particular) talk much of creativity. Vast sums of money are spent in training teachers and other workers to teach others to be 'creative'. Perhaps we should ask ourselves if all this activity to achieve maximum creativity in the greatest number of people is practical and possible and therefore useful, or impractical, impossible and therefore, in the main, useless. Would we perhaps be more effective in our so-called art teaching and eventually as industrial producers right across the board, if we answered this question clearly?

There are many kinds of creativity but, broadly speaking, they can be divided into two categories: creativity in the biblical sense, 'In the beginning God created the heaven and the earth . . .' out of unimaginable nothingness. This kind of creation involves the replacement of nothing by something quite independent from that which has gone before; it implies originating in the full sense with unpredictable results.

At the other end of the creative scale is the imaginative effort required to fit together a jigsaw puzzle which is lying all jumbled up in its box. The outcome of this effort is totally predictable.

In between these two extremes there is an endless variety of degrees of creativity, from the conception of the double helix to explain the construction of RNA, to man-made fibres and plastics cobbled together out of existing and familiar molecules, to the juxtaposition of fur on bare flesh to titillate the jaded palates of haute couture customers and to the combination of drugs, hair, dirt, violence and foul language to make life tolerable to hippies of various persuasions.

There is no doubt that, within these parameters, everybody can be creative just as there is no doubt that, if we misuse the term creative by allowing it to be attached to all human activity, then it will lose its meaning altogether. This is why it seems to me important to understand first what human creativity really means and what the true creativity content (factor) of various activities is; next it is important, I think, to realise that not all people can be creative in the full sense of the word; and, finally, it is important to try and identify talent for what it is in an educational context and equally necessary to adjust educational ends and means to the realities of the creative abilities of the pupils. In other words, I do not think that much good could come from efforts to breed a generation of Einsteins in the physics department of a university (even if a formula for breeding Einsteins existed); on the other hand, if average ability is factually established and a training appropriate to it is given, the chances are that many usefully occupied physicists will be produced most of the time and that somebody in the Einstein mould might pop up once in a hundred years.

It seems to me a pity that art and crafts training at all levels is based on the assumption that most pupils can be fully creative and that, with some help, they could, most of them, become Picassos; or that even if they do not become Picassos, they should experience the thrill of creativity just as Picasso must have done.

It is also sad that our educational system seems to be run on the principle that in art there is either creativity or nothing. It is all right to learn facts without attempting invention in science, or in literature or in music; in art and crafts it *has* to be invention and creation.

I should like to suggest that one of the common factors in all human activity is the need to absorb accumulated experience and to apply it to contemporary situations. This is an activity which can be experienced with great benefit by the vast majority of people and is indispensable for progress in any sense. It is an activity which has been somewhat neglected in regard to art and crafts and in regard to the natural and man-made environment which we seem to take for granted. It seems to me that school curricula, which take into account the fact of pupils' very limited ability to create but comparatively limitless ability to enjoy the work of creators, would be much more fun and immediately productive from the pupils' point of view and would serve them much better later when they become consumers and can influence the producers.

X

The source of all human actions has always been of great interest to philosophers and scientists. The triggering mechanism for the creative impulse has been of special interest because the ability to create is so rare and so important to human progress. Since it is the purpose of this work to suggest that, in art and crafts classes, the emphasis be shifted from teaching children to be creative to teaching them to see and to understand the work of grown-ups who have been creative, it may help to have an up-to-date idea of why some people create and others do not and what physiological processes are involved in creation and in activities which are akin to creation.

H. J. Campbell, in his recently published book *The Pleasure Areas*, has put forward a theory which explains most human activities in neuro-physiological terms and so makes it possible to trace the motives behind our actions.

Campbell's theory is that all animals, including the human animal, are pleasure-seekers first and last and that the moment pleasure-seeking ceases, life itself ceases. To seek pleasure is different from enjoying any given experience. It implies a sequence of enjoyment, satiation and the search for new pleasures. The experience of pleasure is something physical which, according to Campbell, can be measured exactly by the amount of electrically active impulses which are generated in a certain part of the brain which Campbell calls the Pleasure Areas. Through Campbell's laboratory experiments it became demonstrably obvious that pleasure is not just a feeling. It is as much a material fact as the contraction of a muscle or a beat of the heart. Pleasure is simply the name we give to the subjective feeling we experience when our limbic areas are electrically active. The simplest way to pleasure for most people is through the senses because it is by means of the senses that we detect events in the first place. In normal animals, including man, the pleasure areas, deep inside the brain, are activated when the sense organs, out on the periphery of the body, the eyes, the skin, the nose, the ears and the tongue, are stimulated.

Many people can obtain pleasure without the use of the sense organs at all, or by means in which the sense organs play only an ancillary role. Sense organs play only an auxiliary part, while the mind plays a major part, for example, in the pleasure generated by day-dreaming, classical music, painting, sculpture, architecture, writing and reading. This, says Campbell, is where man differs from animals: in the use of the brain in respect to the activation of the pleasure areas. Laboratory experiments in neuro-anatomy and neuro-physiology have demonstrated that nerve impulses pass from the very highest regions of the brain, the cerebral cortex, down to the lower, older areas including the limbic system. The same downward flow of nerve impulses from the higher regions to the limbic system is the source of human pleasure. A human being is able to evoke electrical activity in the limbic pleasure areas by processes occurring in the thinking regions of the brain. This means of obtaining pleasure seems to be totally absent from the lower animals.

People, it seems, can be divided into three main groups, which inevitably overlap and in which we are dealing with their predominant mode of achieving pleasure. There are the 'feelers', the 'thinking-feelers' and the 'thinkers'.

The 'feelers' are those who, throughout most of their lives, obtain pleasure mainly and directly through one of their five senses and only to a small extent from sensations which are derived from any form of thinking or intellectual activity. The 'thinking-feelers' are people who derive their pleasure predominantly from processes of thought (which may have been triggered off by one of their senses) and which are then converted into material form. Then there is the small group of people who derive most of their pleasure from the operation of the higher mental processes.

The 'feeling' people do not really take advantage of the advanced evolution of their own brains and, while apparently engaging in human behaviour, are in fact utilising mainly the sub-human neural mechanisms. The 'thinking-feelers', having brought the sensory mechanisms under the creative dominance of the higher regions of the brain, may start their pleasure cycle by first having their senses stimulated from outside. These sensations are then processed by the higher regions of the brain which either by 'understanding' or by conceiving new thought generate the impulses identified as pleasure. It is necessary for such people to use their senses in stimulating forms in order to communicate their ideas. Mathematicians, philosophers and other 'pure thinkers' have relegated the pleasure to be derived through their sensory receptors to the essentials of life and depend for real pleasure on the neocortex.

I have just said in regard to the 'thinking-feeling' people that they are assumed to have 'brought the sensory mechanism under the creative dominance of the higher regions'. What I suppose I meant was that these people are not born with their sensory mechanism under the dominance of the higher regions (which themselves are, at birth, as yet underdeveloped) but that the higher regions of the brain obtain this dominance with the passage of time, with the investment of work and with the maturing of the individual. Put in yet a different way, even the people in the second and third categories, the 'thinking-feelers' and the 'thinkers', when they were children, were most unlikely to have had their sensory mechanisms under the dominance of the higher regions of the brain; as children they obtained their pleasures direct from their sensory mechanisms just as the 'feeling' people did and still do. One would suppose this to be a fact of life and there is little that the child or the teacher can do about it. A child will therefore be quite simply looking for pleasure through its senses. In art and craft classes this means that the child will respond much more readily to the stimulation which the senses can receive from sights, sounds and textures than to the stimulation and pleasure received and generated from the abstract mental effort required for creation. Maths is a difficult subject for most children because it is difficult to introduce any sense-tickling element into it. Yet what children are expected to do in maths classes is to understand rather than to create. To take a sense-satisfying subject like art and craft – something which children are ready for – and transmute it into an intellectual activity which creation implies and which is something that children are not ready for, is probably pointless.

The metamorphosis of the young 'feeler' into the mature 'pure thinker', which does not necessarily mean 'creator', is, it seems to me, if one might be allowed to generalise from the earlier description of a hypothetical experience in seeing, feeling, doing and abstracting, not only true in the long term and spread over a life span, but is also encapsulated in each visual experience which is allowed to run its natural course. Even the 'thinker' is sometimes, and the 'thinking-feeler' almost all of the time, a 'feeler' at the beginning of a creative cycle. The significance of these two points is that they make it clear that 'feeling' is indispensable in 'thinking-feeling' and perhaps even in pure 'thinking' and that it must be encouraged in children if they are expected to develop and to grow. It should perhaps be encouraged even in adults if their senses are not to atrophy. Therefore there is nothing wrong, when teaching children, to put the emphasis on feeling and reacting rather than on abstract thinking, because children are capable of the one but not of the other.

It may perhaps be useful to summarise Campbell's theory:

1. It is the requirements in all vertebrate animals that the pleasure areas of the brain shall be maintained in a state of electrical activity.
2. In sub-human animals the state of activity of the pleasure areas is maintained by nerve impulses originating from stimulation of peripheral sensory receptors.
3. In human animals, uniquely, there is a second source of activity in the pleasure areas, namely by nerve impulses originating from the thinking areas of the brain.
4. In all humans, behaviour is directed at one of three means of obtaining pleasure – (a) through the senses; (b) by using a combination of the mind and the senses; and (c) by using the thinking regions as the sole source of pleasure.
5. Satiation is an essential feature of pleasure seeking. In order to live, animals, including the human animal, have to vary their pleasures. The uninterrupted enjoyment of any single activity would result in death.

It is difficult to over-emphasise – in the educational context – the fact that, to be successful, teaching must take account of the pleasure-seeking basis of life and that children obtain pleasure through the senses rather than through the mind.

It becomes obvious then that, amongst other things, healthy, normal children cannot possibly like the disciplines of school. As pleasure-seeking animals they derive their pleasures directly through their senses and not through the neocortex which even the simplest lessons in English or maths must utilise. As pleasure-seekers they are looking for immediate sensory stimulation and teaching them abstract subjects which do not first stimulate one of their senses has to be made extraordinarily attractive to give them pleasure. At best, that is when they start off by enjoying a period of an abstract subject, they soon get bored and look for fresh sources of pleasure. At worst, when they are uninterested in the subject or find the discipline too irksome, they positively refuse to cooperate with the teacher and to learn. This may be why what we call art classes are so popular at primary-school level: colour and texture offer direct sensory stimulation: the eyes behold, the fingers touch and this causes pleasure in straightforward physical terms. The children are interested in having their senses stimulated and incidentally and simultaneously they also gather information and gain experience and possibly some skills; they learn, if 'art' is imaginatively linked to a number of other subjects. From the teacher's point of view they are easier to control because they are busy but mainly because they are happy and satisfied, experiencing pleasure. This explanation of the popularity of art at school clearly excludes the possibility of children being creative in class or being taught to be creative.

We should therefore note and recognise what children are capable of and try to meet their natural demands and develop their natural abilities within their limitations. To make them perform feats beyond their capabilities is pointless and could turn a pleasurable activity into a painful one (when the child finds that it cannot perform the feat). We are happy to teach children to play the piano and a very few of them play it prodigiously well; it is accepted that they are, as a rule, not taught to compose or to interpret a work or to add variations of their own. (There is also the sad sight of children who are made to play a musical instrument and spend thousands of hours practising. Playing an instrument is then transformed into a social asset. Many of these young 'musicians' have no idea of or liking for music in general, never go to concerts or listen to music because they enjoy it. They have little knowledge of the work of even the most well-known composers – except for those pieces which they have practised.) Children are taught to read and to write, often a painful experience for them; and they are often set to copy slices of literature directly or in the form of précis; or they are asked to narrate their personal experiences (which is another form of copying). We do not attempt to teach them to create literature or poetry; we try to familiarise them with literature so that they can recognise the work of classic and modern writers and so that they may enjoy the individuality and the talents of all of them. All this means that we ask them to re-create, not to create. Similarly, children can be taught to see what is around them, to touch such parts of their environment which invite contact, to draw more or less well – which means to copy, to paint (to copy more thoroughly) and to explore in many different ways the physical qualities of their environment and the objects in it. They can be taught to narrate graphically; they can even be taught to understand and appreciate the creativity of others in the visual media; and they will do all these things well because all this is direct sensory stimulation and gives immediate pleasure; but they cannot be taught to originate and to create; to try to teach them this would appear to me to be so much wasted effort and positively harmful because it is bound to be frustrating and off putting. To shape and sharpen children's understanding in regard to what they see, on the other hand, is just as necessary as the development and growth of their other faculties.

It seems reasonable to work on the assumption that children can become acquainted with and can reflect the visible and can do so and gain pleasure while doing it, but they cannot 'make the invisible visible' without strain and without losing that sense of pleasure – if at all.

Indeed it is doubtful if creativity can be taught – to children or to adults. It can be explained up to a point but it cannot, I think, be taught in the usual way even by genuinely creative people. (Probably the nearest we come to teaching creativity is by example and apprenticeship.) Observation, no matter how accurate and useful, and the acknowledgment of a visual fact and its reproduction is not creation and can be learned. Creation is much more difficult to define and probably cannot be learned.

Brian Allison, in *After Hornsey*,* puts it like this:

> It is possible to look at the work of eleven-year-olds at the school where the work is done and to predict exactly what that work is going to look like when the child is in the sixth form. The same is true of college.

* *After Hornsey, Reading in Art and Design Education.* Edited by David Warren Piper (Davis-Poynter, 1973).

It is clear therefore that art forms at pupil and student level are mainly derivative and we should distinguish between making art and creating art. It is one of the false premises on which art education in this country rests that creating art – being original – is possible for all. It isn't. It is only possible for a gifted few.

Making art is certainly possible for almost everybody but to make art serves a different function from creating it and to recognise this means changing the way we teach art in schools and anywhere else. We seem to believe that art teaching must be connected with expression (of feelings, of thought, of moods). This is possible for the very few for whom the visual arts are an appropriate vehicle of expression but for most children (or adults) this is quite impossible.

Our art education seems to have taken the artist as its model and this is quite wrong; it would be more appropriate to choose the sensitively responding and discriminating user as a model. We seem to believe that participation in art activities on a practical level carries with it – automatically – a development of aesthetic sensibilities, discrimination and enjoyment. An examination of the pupils from two secondary schools whose practical activities were indistinguishable showed that such a transfer from activity to judgment does not take place. Only one of the teachers in the two primary schools which the children had previously attended had the clear objective of teaching aesthetic sensitivity and discrimination; the children from this school showed at fifth-form level a better aesthetic judgment than children in the first year. The children from the other school, where the teachers did not have such an objective but had simply engaged them in what they thought was creative art, had no different judgment in the fifth form from the first-year children in the school.

Education, Allison thinks, has two functions: the transmission of culture – the acquisition of skills which will enable students to profit from the contributions of those who have gone before – and the equipping of students to a more original contribution of their own: Instructional and Expressive objective. But they must be kept apart and taught, if at all, separately.

It seems fair to conclude that the teaching of visual creativity cannot and should not be part of any sensible and useful general curriculum at school, but that observation, investigation and assessment of our natural and man-made environments and objects can and should be; that appreciation of visual matters and of creativity should indeed, and urgently, become part of the curricula in our schools. There is something faintly monstrous in seeing otherwise well educated people who have remained visually retarded and illiterate; just as there would be something faintly revolting in seeing a Château-Lafite Rothschild being drunk out of the bottle, between mouthfuls of fish and chips and HP sauce, lifted with the fingers out of the *News Of The World*; it is necessary for the minimally educated person to be able to assess the usefulness, the physical qualities, value, rôle and the aesthetic qualities of the artefacts in daily life; and specially trained teachers are required for this purpose.

XI

The position today and the changes one might, in the normal course of events, expect to see in the next few years in attitudes to art and crafts in our primary schools, can be seen in a report, published in 1967 by the Secretary of State for Scotland on art and craft in the primary school. (If one discounts the myth of the excellence of Scottish education compared to the rest of the country – as one's daily experiences have given one every reason to do – the report and its implications can be reasonably applied to Britain as a whole.)

The introduction to the report says:

For long, art and crafts occupied a peripheral position in primary education. Teachers, largely because of their own experiences as pupils, regarded them as skills, useful at time and pleasure-giving but frequently to be restricted or even dispensed with in the face of other pressures. In recent years there has been a general acceptance of their value as a means of expression and a form of communication. In consequence, there has been an upsurge of interest which can be readily recognised in the increasing number of activities and in the greater variety of media now being used.

The evidence points also to a growing body of teachers who find little satisfaction in either the arid formalism of earlier methods or of the more recent cult of 'free expression', which often fails to develop the pupils' potential because of the lack of clear objectives. These teachers recognise the need for a fuller appreciation of the basic aims of education through art and crafts to help them in their task of providing both stimulation and inspiration. Many teachers, however, for various reasons, lack confidence in their own ability to develop the educational opportunities offered by art and crafts; and much of the development that has taken place has stemmed, not from a full understanding of the aims, but from the enthusiasm of individual teachers. Admirable though these qualities are, they provide only a partial base from which to guide the visual and aesthetic education of children.

And further along:

Picture-making plays by far the largest part in children's art activities. It is a medium through which they express what they perceive and discover and in helping them to be aware of themselves and their relationship to the people and things around them it may take a highly personal form. This expressive content and its refreshing vitality, which are common at the early stages, generally diminish in the later years where there has been a tendency for children's artistic experiences to be curtailed and for more importance to be attached to technical skill, neatness and representational accuracy.

Firmly convinced that they are encouraging 'free expression' teachers often suggest themes for painting and modelling with little attempt to motivate thought or to evoke personal reactions in their pupils. When pupils are left entirely to their own devices most tend to repeat subjects and symbols in which they have confidence, but which seldom have meaningful significance to them. They cannot be expected to grow in visual awareness unless they are guided. The teacher has a vital rôle in stimulating the pupils as individuals to see, think and to feel. Her contribution, however, must be carefully balanced according to the needs of each pupil. Art and crafts emphasise the uniqueness of the individual and some pupils' experience, imagination and sensibility may require little of the teacher; others with meagre resources, may need all the encouragement and careful guidance which she can give; and for them art and crafts may be the most meaningful approach to learning.

In the view of the report, *Art and Crafts in the Primary School**, much has been done to link art and crafts activities to spoken and written English, environmental studies and other aspects of the school curriculum; looking at staff attitudes it comments:

The most encouraging signs of progress and development have undoubtedly been seen at the earlier stages, particularly in infant classes, where many teachers have displayed an awareness of the value of the activities as a means by which their pupils can relive experiences and express and communicate personal feelings and ideas. Pupils are given fewer opportunities of this nature at the middle and later stages and in the case of the oldest pupils the pressures of internal and external examinations have sometimes been used as an excuse for restricting or even eliminating art and crafts, if not for a whole year, at least for a part of it.

*An extract from the Report of the Secretary of State for Scotland on Education in Scotland in 1967.

There is no stage more urgently in need of objective planning than the last two years of the primary school where the pupils require wide and full artistic experiences. Few teachers of these pupils have had an adequate art education themselves, either at school or during training, to enable them to meet this need. As a result of the provision now made by some of the colleges of education it is confidently expected that more young teachers will reach primary schools with a fuller understanding and appreciation of the aims of art and crafts activities in relation to children's learning experiences.

The report concludes:

The opportunities in primary schools for developing education through art and crafts are greater today than at any time in the past. There is a growing awareness of art and crafts as a necessary part of the full and balanced development of the child, an attitude which has been influenced by current educational thinking about all primary school work. Improved accommodation and a greater variety of media have encouraged experimentation and development; and not only is there evidence of a more liberal training of young teachers but it is also recognised that there should be a more consistent policy of in-service training. As in all other aspects of education, success or failure depends upon the qualities of head teachers and their assistants. In addition to their educational conviction and their understanding of the education which can be provided through art and crafts, they require to have a degree of aesthetic sensibility and a knowledge of the materials and equipment which they use. The teacher who may feel inadequate aesthetically has still a valuable contribution to make. She can develop perception, stimulate imaginative thinking and encourage personal expression, qualities which are the very essence of learning and living. Aesthetic sensibility, however, is much more fundamental and its roots reach back through primary and secondary education to one's earliest experiences. Only if a rich art education is provided at both primary and secondary levels will the child's sensibilities be given an opportunity to develop to the full.

This is so obviously a well-meaning report that it may seem pedantic to talk of it in critical terms. Yet one wonders if it is quite as clear in its thinking as it could and as free of clichés as it certainly should be. Take the form of words used throughout: Art and Crafts; joined together as if they were the kind of Siamese twins which it would be dangerous to separate. Art and crafts have not a great deal in common and the talents and training which make a good craftsman could be a hindrance to the artist. It does not seem to have occurred to the drafters of the report that art may have no more in common with crafts than with English and should therefore not be automatically lumped together with it but be treated as a separate subject.

Next, art, by definition, is not a natural phenomenon or an accident but an idea which is first conceived then realised and made palpable in a controlled fashion. The most beautiful sunset is a natural phenomenon, not art; the simplest line drawing by a person with a trained eye and skilled hand which expresses a personal point of view is art and not a natural phenomenon. To call the 'art-work' of children 'art' in the normal sense is therefore not helpful to our understanding of what the children are actually doing and in regard to what the teachers should be doing.

The report emphasises the expressive nature of children's art but is not quite as clear as it could be about its exploratory nature, about the fact that, among other things, children engage in what looks like art because they are simply trying to establish just what, say, crayons, water colours, poster paint or oil paint look like on ordinary paper, on newspaper and on cartridge paper; how clay feels and behaves when pummelled, rolled, pressed or coiled; how different natural and man-made fibres feel when spun, woven or knitted; how differently they behave when print is applied to them and so on and on. It would be boring and impossible for children simply to try and establish the physical facts in regard to the materials in their environment in a methodical, abstract, scientific manner, for while they are curious above everything else they are also impulsive and lack the necessary motivation, knowledge and self-control to conduct purely factual enquiries. They will always tend to occupy themselves with what gives immediate pleasure rather than with what will benefit them (give pleasure) in the long run. One important aspect of growing up is the slow shifting of emphasis from instant satisfaction to postponed satisfaction and this is something that children cannot be expected fully to realise and live up to.

Children in primary and secondary schools are quite simply looking for immediate pleasure. Playing with colours, textures and various raw materials gives immediate visual and tactile satisfaction and pleasure and will, therefore, be willingly taken up by most children as an end in itself. As far as the pleasure derived is concerned, this is no different than playing with toys. Fortunately such play has beneficial, long-term aspects and effects which teachers can channel into serving other, worthwhile ends. It is these secondary aspects and their comparatively long-term ends which are well – if perhaps a little naïvely – spelt out in the report; for both the quality and the quantity of such instruction as is available could not by any stretch of the imagination be expected to produce the results which are confidently forecast in the report.

It is important to understand the basic nature of children's art-activity, which is curiosity-play – pleasure-seeking, as distinct from its subsidiary aspects. Only then can the activity itself take its proper place in the time-table, only then can teachers be suitably trained for its supervision and only then will this healthy, spontaneous activity retain its innocence and force, and become the foundation on which the fully realised, well educated, sensitive and discriminating individual can build and develop. It seems strange that it does not occur to us that, when children play 'ring a ring of roses', they are actually doing ballet or that, when they build houses with wooden bricks or with 'Lego', they are doing architecture, that when they crawl under the table and play houses, they are doing domestic science. Yet when they do at school exactly what they would do at home in regard to drawing and painting, we call it art. It is perhaps our constant and long accepted linking of 'art and crafts' that causes this confusion; for while there is obviously a certain tenuous connection between these two activities the differences between them are far more fundamental. Perhaps in recognition of this we should separate the two and never again mention them in the same breath.

The report, good as it is as far as it goes, makes no mention of the importance of an analytical or critical approach to the visual world and especially to man-made objects, the artefacts in the children's environment. By saying, as it does, that 'there is no stage more urgently in need of objective planning than the last two years of primary school where pupils require wide and full artistic experiences' it tends to obscure the true nature of 'art and crafts' training at this stage in a child's life and seems to ignore the much greater need for the objective planning of the last two years at the secondary schools – especially now that the school-leaving age has been extended. The problems at this later stage in our educational system have been and are well known and the need for a new approach is quite desperate. This is the point where we launch into the world bored, resentful and misinformed wage-earners who mostly lack any sense of values, instead of interested, knowledgeable and eager young people with open minds and open eyes, who are receptive to ideas even if these emanate from people older than themselves.

There seem to be three distinct motivational sources for art and craft activities at primary level. First, the satisfaction of natural curiosity in general – the familiarisation with environmental objects: what happens if one mixes paint and water and applies the mixture to paper? Second, the satisfaction of a natural curiosity about oneself, one's relation to the environment and to other people – the familiarisation with oneself, including its expressive aspect. (One fed the squirrels in the park. How did one do it; what did the squirrels look like; and how did one look feeding them?)

Third, the satisfaction of a natural curiosity about other people, about their work and about the relevance of that work to oneself – the familiarisation with others and with analytical and critical aspect of others' work and of one's own in relation to them: (Johnny's father makes chairs and one can buy them in the shop for ten pounds. How does he make them? Why does one have to pay ten pounds for them? What is special about them? What would one do with such a chair if one bought one? And so on.)

While it is possible, as the report envisages, to aim at training primary-school teachers to meet, better than they do now, the demands which the first two aspects of art and crafts activity entail, it is difficult to see how the same teachers can also meet demands of the third kind which entail a different kind of training and different personal resources from those which teachers are normally given and possess. They imply training in the comparative study of industrial output and the constant updating of such information.

Art and crafts and environmental appreciation in the correct sense of the word should start at primary level. This should not be concerned with creating in the true sense, with the making of original things by the pupils. It should really mean 'getting acquainted with art and crafts', that is, with art as created by artists and with crafts as made by craftsmen. It should be the responsibility of the specialist, or visiting teacher, who should work hand in hand with the resident teacher. The former could have two clearly defined responsibilities: to familiarise pupils, to the extent to which their intellectual equipment allows, with the work of craftsmen and artists, past and present, to plant into their minds the idea of discipline, training, effort and organisation that converts the products of instinctive activity into works of art or crafts; and to initiate them into the habit of an appreciative, an analytical and a critical approach in relation to their environment with special reference to the artefacts it contains. (Some teachers clearly feel that a critical approach to the children's home environment might cause problems. Obviously it is important that no embarrassments or resentments should be caused as a result of teaching this subject in this way. The tact required can, I feel certain, be taken for granted in most teachers.) Exploration, familiarisation and self-expression should be encouraged by the resident teacher but should be known under these headings and not under Art and Crafts.

The teacher who could perform the function of relating the work of artists, designers and craftsmen to day-to-day living would, of necessity, have to be of a new kind: trained not as a creative artist but as a person with high visual awareness in art and in design who could make children and young people share his visual experiences and could encourage them to develop their own points of view in regard to what they see.

XII

In order to establish and understand what goes on in our secondary schools, who the teachers are, what they think and what they teach in regard to art and crafts, I prepared a list of questions and arranged twenty interviews with teachers. The questions were based on my own ideas of the subject of art and crafts and were aimed at establishing some facts. I wanted to satisfy my curiosity so as to make it possible for reasonably well informed conclusions to be reached. The interviews were conducted by Gillian Colledge who also taped and typed them, were spread all over the country and took about four months to complete. Gillian and I met every few weeks for a discussion and so that certain adjustments could, if necessary, be made to the sequence of the questions and to the questions themselves.*

Among the things which struck me immediately upon reading the interviews was the wide gap which exists between the everyday realities at school and the comparative lack of reality in the annual or bi-annual reports which keep coming from committees and other bodies. A summary of one such report and certain comments on it are to be found in Appendix C. The report is that of the Working Party on Art in Secondary Schools, prepared for the Scottish Education Department and published in 1971. Typical of its kind, it differs little from other similar reports published in Britain since the Fifties: well intentioned if somewhat avuncular, plausible but rather shallow, covering much ground yet leaving art education pretty much where it found it: the least important and the least serious subject in the curriculum. In an industrial country which exists by exporting its industrial products this is inexplicable.

Not as inexplicable though as some other striking impressions which come through on reading the interviews side by side with one's daily papers. For many of the teachers said that there simply

*Photocopies of these interviews are available on loan, on request to the author.

is not enough time to increase the amount of instruction in art which the pupils receive; they said that there is not enough time even to cover adequately the present curriculum. Yet, according to the papers, the recently introduced legislation for the extension of the school-leaving age is a total disaster because no one apparently knows what to do with the time which has suddenly become available for extra instruction.

The extension of the school-leaving age from fifteen to sixteen is an important step forward in the direction of bringing Britain into line with other developed, Western, industrialised countries and so removing, albeit in the long term, some of our self-imposed handicaps vis-à-vis our competitors in the world markets. For many years now, everybody in Britain has agreed that another three terms at school were desirable; but not all governments had the courage to put this accepted view into practice. All of us, however, those with and those without courage, have still to realise that in this case good intentions are not enough because those fifteen-year-olds who would normally leave school at that age but are now kept for another year – and who are the majority of pupils – need very special inducements and attention-catching devices if their minds are to be stretched and broadened for another year. To give them more of the same instead is completely short-sighted, self-defeating and wasteful.

Our interviews confirmed the suspicion that a great opportunity for real improvement in our system of education is being lost because of certain mock battles and uncertain victories which have been fought and won these past few years. We do not seem to have worked out at all carefully exactly what we want to do with this extra year at school which our children now have. We have all been so busy fighting verbal battles that at no time have we bothered to understand the fifteen-year-old's mentality which is impatient with school, itching to leave it, its books, its teachers and its younger children, to become part of the grown-up, 'free', wage-earning world of 'work' – of putting into work as little as one can get away with – and of treating money which has thus been 'earned' almost as if it did not matter at all. It is indeed difficult to imagine why a youngster awaiting such a release and striving for such a life should want to spend another year with his teachers whom he does not respect or with his books which he does not like.

It is not too difficult to imagine, however, that young people who are not academically inclined would be quite interested in finding out more about the world into which they intend shortly to plunge. They could conceivably be interested in finding out about the thousand and one things which they would like to possess from clothes to cars and would, very likely, listen with a fair amount of attention to the detailed description of a good, worthwhile holiday compared to one where they would be exploited and overcharged; or to the comparison of two plays or films where they have to make a choice to see one; or to a comparison of pop records. They might also enjoy a certain amount of travel: in a country as rich in its architectural heritage as Britain, excursions in the off-peak seasons, tailored to their special requirements could prove well worthwhile. There are also the permanent and other exhibitions or the institutions which they might like to learn about and see.

The magnitude of our failure to realise the potential for good which this additional year at school suddenly made possible can be appreciated only when the aridity of this last year in our secondary schools is considered and is compared to what could be.

The interviews also show pretty clearly that, perhaps more than in most other professions, in the teaching of art it is not mainly the challenge that attracts but the safety of tenure and the lack of commercialism and competitiveness. This is perhaps inevitable to some degree in all teaching and a good thing too; yet one cannot help wondering if the results would not be better if the requirements and the status of art teaching were to be lifted. Both the education authorities and parents allocate a position to visual art well below the other subjects and even below other art subjects such as music. It is possible that the lack of hard and fast, easily recognised rules in regard to art makes parents nervous of it as a breadwinning tool and the authorities' attitude may reflect this. It is evident from the teachers' remarks that art is not recognised for the overriding subject that it is and into which it increasingly evolves in the course of the life span of the individual.

The way the teachers expressed themselves during the interviews shows quite well that there is, as yet, no clear differentiation in their minds between the tasks of discovering creative talent and encouraging it and the fostering of an all-round sensibility to everything seen in pupils where talent does not exist. Like the rest of us, many teachers seem to think vaguely that everybody can be creative, to a large extent at any rate, and that no effort should be spared in art classes to enable all pupils to be creative. Most teachers also feel that the task of increasing visual sensitivity is important but, if anything, even more difficult than making pupils creative. They think this is so not only because there does not seem to be an agreed system of training for visual sensitivity but also because of the many other counter-influences which have an effect on the pupils' outlook in this respect and which counteract the school influences. The fact which emerges quite plainly from the interviews is that little or no training in visual awareness and sensitivity as such takes place at elementary and secondary levels.

One would have expected the Design Council to come into its own in this sphere; to suggest new principles, to guide towards new approaches and, above all, to help with up-to-date, practical visual aids. Unfortunately, to most of the teachers the Design Council is barely a name.

Most young people who go to art schools prefer to think of themselves as potentially creative artists rather than as potential teachers of seeing. More than in most professions and notwithstanding the Foundation Course which is intended to help sort out people and things, a very large proportion of students go in for a training which would suit them for the rôle of the artist but they end up working as teachers. One is particularly concerned with the point that, even if for a number of reasons this is unavoidable, teachers ought to be trained quite differently from artists since their talents are and their ambitions ought to be different.

One of the more interesting things which transpire from the interviews is the difference in attitude, in the teaching profession as a whole, to those who intend to take up art as a profession and those who intend to teach art to others and so make it a meaning-giver to ordinary living. The teaching profession accepts and respects the first but seems to have little time for the second. It is not generally realised by the teachers and by the taught that to be civilised one has to cultivate the visual senses in general and not only the 'fine art senses'.

Another feature of teachers' thinking is the strongly felt obligation to help pupils pass examinations with the highest possible marks. This is natural and has the expected sequel that since the pupil can sit an examination in drawing, drawing is taught; but since there are no examinations in visual appreciation, visual appreciation is not taught. Paradoxically most interviewed teachers felt that their aim at school was to produce the rounded, visually aware person rather than draughtsmen but were somewhat vague about the means they used to achieve this end.

One wonders if it is really beyond human ingenuity to devise examination papers for visual appreciation, judgment and sensitivity?

The feeling of hopeless despair is also quite noticeable in some of the interviews. As one teacher put it:

> One feels that if a highly specialised public school with relatively many more facilities and fairly ideal working conditions, small groups and so on, can't turn out an educated boy who will be the natural patron of good design, whatever that may mean, in every field, then there isn't much hope for anybody and I don't think we are turning out such boys.

Another teacher said:

> A large percentage of the boys have really done absolutely nothing or have been subtly indoctrinated with the idea that art is something you give up as serious subjects move in.

Or:

> They (the boys) feel that everything to do with art is a bit of a joke and they are absolutely non-conversant with the simplest points about design. To pick up a book, for instance and actually be slightly interested in the way the design of the jacket is actually done or to feel that it is an object to look at as well as to read – they just don't see this.

The reason behind the career-selection of many teachers is perhaps contained in the well known story by Somerset Maugham, of which I have always been rather fond. It is called 'The Verger' and concerns Albert Edward Foreman, the Verger of St Peter's, Neville Square, a fashionable church in a good neighbourhood, who was dismissed from his post when it was found that he was illiterate.

Foreman, dejected and forlorn, was walking home and felt like a cigarette. In the long street in which he found himself, with all sorts of shops, there was not a single one where he could buy a packet of cigarettes. This gave him the idea of opening a tobacconist's in that street and, when he prospered there, he went out in search of similar streets – long and tobacco-less – and over the years established a whole chain of them, all of which did remarkably well and from all of which he collected the takings personally to deposit them in his bank.

On one such visit to his bank, the manager suggested certain investments to him and said in conclusion: '"We'll do everything. All you have to do next time you come in is just to sign all the transfers."

"I could do that all right," said Albert uncertainly. "But 'ow should I know what I was signin'?"' And so it came out that Mr Foreman could, apart from signing his name, neither read nor write.

The manager, unbelieving, exclaimed: '"Do you mean to say that you have built up this important business . . . without being able to read or write? Good God, man, what would you be now if you had been able to?"

"I can tell you that, sir," said Mr Foreman, a little smile on his still aristocratic features. "I'd be Verger of St Peter's, Neville Square."'

Most art teachers might well have been happier and in other careers without their art education. For, on reading the interviews with the teachers, I wonder how many disappointed art teachers today would be successful in other careers had they not, early on, had such good qualifications, on paper, for art teaching – had they indeed gone into other careers. Certainly I know a number who left art teaching and have made a considerable success of their various businesses. The fact seems to be that the majority of art teachers did not set out to become art teachers but chose teaching because they found it difficult at the time to fight circumstances; they had most of the qualifications for teaching art and, with a little extra trouble, acquired the few missing ones. A better formula for thousands of round pegs in square holes is difficult to imagine.

The sense of frustration and the futility of trying to set higher visual standards than those which prevail at home and in the streets, on television and in the papers and magazines is also quite evident from the teachers' remarks, coupled to complete puzzlement about a way out of this head-on clash between the existing environment outside school and the theories they try to propound inside it.

Summing up, a fairly clear picture of individual enthusiasm and of some good practical work emerges from the interviews; but there is also a distinct feeling that many of our secondary-school teachers go into art teaching, not as a vocation but as one of the choices, well down the list of their preferences, with which life in the raw has presented them. We also sense that little has fundamentally changed in the way art is taught: it is based on an article of faith in creativity as such and in the ability of each individual to create to a sufficiently large extent to make it worthwhile spending most of the time available for the teaching of art and crafts in teaching creativity. It is assumed that the 'spin-off' from this activity will be sufficient to see the pupil through life as far as general visual awareness and sensitivity are concerned. Above all there seems to be no true realisation at any point in our educational hierarchy that visual literacy and general sensitivity are exactly as important to the individual and to a developed and highly industrialised country as literacy and numeracy are to the underdeveloped and backward countries of the world; and that this subject demands to be treated as an important one in its own right; that it deserves a full and well thought-out curriculum; that it should almost become the only subject taught in the last year at school; and that many extra teachers are needed for it and that those teachers will have to be of a new kind.

XIII

Art schools have three functions: to produce artists, of whom the community needs very few compared to its own numbers; to produce designers for the crafts and for industry, of whom many more are needed by any society which earns its living in industry; and to produce teachers of art and crafts, of whom many more are needed than designers or artists.

What strikes one immediately, on looking into what goes on in our art schools and in our schools of art and crafts today, is that there is much debate about the education of the designers and artists but almost none about the education of the art and crafts teachers and of course none about the education of teachers of looking and seeing. At a time when we need at least ten times as many teachers of art and crafts and of seeing as we need designers and artists, this is quite remarkable even if this can be explained by the fact that the influence of the teacher of seeing on our economic well-being is not yet appreciated.

(The current debate on the designer is not without interest. It is concerned, on the face of it, with the desirability of training specialists as against generalists and with the 'network' system, the 'cafeteria' system and the 'grid' system of education compared with the more old-fashioned systems. A summary of this debate is given in Appendix D.)

It seems, too, that it is taken for granted by the authorities that, even though most of the students who attend art schools will end up as teachers, it is generally accepted that they have to be trained for different careers and not for teaching; that there is no need to rethink the training given to teachers and that it is good to leave it as it essentially was twenty-five years ago; and that the rôle of the art and crafts teacher will remain almost completely unchanged even though the world is changing rapidly and at an accelerating rate.

Our children, about ten million of them at any one time, spend about ten years in primary and secondary schools. In that period most of them get next to no systematic training in visual and environmental appreciation and discrimination. It is no exaggeration to say that where 'art' is taught most – at primary levels – it is the modern equivalent of the sweeties we used to be given as children for good behaviour and for working well at school; and has as much real educational meaning; the art education they receive does the children of today as much lasting good as the sweets of fifty years ago did us. Art in primary and elementary schools has a homeostatic function – it has a pedagogic-survival value (it enables the general programme of teaching to be carried on) and is not of much use in itself for its own sake. All our children will become consumers of design with no more preparation for this rôle than they had by the time they left secondary school and only marginally more than when they entered it; in other words with no preparation at all. This cannot be good enough from a personal, ethical or practical, national point of view.

There are a number of questions which come to mind as soon as one starts thinking about art schools:

How many and what proportion of our children attend them and why?

Do our art schools turn out artists, designers and teachers in a reasonable ratio to the proportion of the students who attend them or is the number of those who put their art school training to practical use disproportionately low?

Should we train artists and designers in more or less the same way, at the same time and in the same classes and as we train teachers?

How many of each shall we need in the future?

Can art be 'taught' in the conventional sense?

Who started the art schools, and when and why, and are the same reasons for the existence of art schools valid today?

Some of these questions are answered in detail in the Appendices. Those that are not might perhaps receive an airing in this chapter.

First views and impressions of art schools can be awe-inspiring. Looking at our better known art schools one has the uneasy feeling that 'In the beginning were the Art Schools'. Created, if not by God himself, then by Olympians not far removed from Him.

The schools of art and crafts have their origins, on the other hand, in the humdrum and recurrent economic crises which the country has repeatedly faced as a result of the Industrial Revolution.

As each generation in turn came face to face with its own problems, it appointed, through Parliament, what it considered an appropriate committee to deal with them. Commissions and Committees have reported on these crises at increasingly frequent intervals over the past one hundred and fifty years and they are still reporting; Governments have acted on their advice (more or less) and they still like to think that they act on it; but the sequence continues relentlessly: crisis, committee, recommendation and some kind of action. The basis of our industrial activity since the industrial revolution has not often been conception, product development, action and exploitation but confrontation (crisis), bewilderment, committee and hurried consideration followed sometimes by some ameliorative action.

It all started off quite beautifully with the Industrial Revolution when invention begat necessity: what was invented in Britain became an obvious necessity in the rest of the world. All would have continued well had British inventiveness continued to lead and had Britain been satisfied in selling inventions rather than in selling the products which the new inventions enabled her to produce in bulk and comparatively cheaply.

The Victorians' problem was how to control the productive machine. Until then there were the fine arts and there were the crafts; they were separate, their functions never overlapped and there was no reason to confuse them. The moment manufacturers realised that manufactured goods had to sell on appearance they also began to suspect that art was a commercial factor and the concept of the designer – something between the artist and the craftsman – began to take shape. The designer's rôle, assigned to him by the Victorians, was to ride and to feed art to the tiger of mechanised production and to keep it purring with production and its owners purring with profits.

In 1832 Sir Robert Peel first raised the matter in the Commons and the result was a committee, and then the establishment of schools of art and design, the opening of numerous art galleries and the protection of invention and design by patent rights. The committee also suggested that on completion of great public buildings, the arts of sculpture and painting might be called in for the embellishment of the architecture. This typifies the Victorians' misunderstanding of art and its relation to crafts. The Victorians were well intentioned but aesthetically primitive.

Prince Albert's rôle as a man of taste and particularly of 'applied taste', is symbolic of the era, of its belief in beauty and of its mistaken notion that everything could be improved, even put right, by the application of a little decoration and design. The paradox of the Victorian age appears now to have been that while they realised that there was some kind of connection between art and crafts, they allowed them just sufficient contact for maximum damage to each. Less contact would have been better – or more.

People then did not realise that, because of the underlying importance of unity in design, decoration and design cannot be separated from or applied to a work; that good design and decoration should start simultaneously with the making of the article. The Victorians barely connected art and design in their minds in this sense and the two were thought of as cousins, as almost different in kind, which had to be produced separately by different people and then brought together. Strangely, there was a built-in advantage in this approach because, had this separation been maintained and had design schools been established which worked closely with industry (and not with art), the chances are that some really sound design development might have taken place from which industry would have benefited greatly; but art and design were for some time thrown together in the schools of art and this did considerable damage to both.

The Victorians were also visually sentimental and their aesthetic sense tended to lushness: flowers ran riot on fabrics, on floors and on firescreens. Structurally unnecessary Greek and Roman columns supported buildings, tables, machines and candles. They said it with flowers or with other vegetation, they said it with tortured, twisted lines and they said it with the bits and pieces of the ancient world. They rarely said it with plain shapes of good proportion, with artefacts which had unity and grace, because they tortured and 'artified' crafts.

One of the handicaps that sentimentality imposes is that it does not encourage logical thought. The logic of the situation in the early part of the Industrial Revolution indicated a special design training for designers which had little in common with the fine art training then available at art schools. But if wrought-iron roses are considered to be necessary and beautiful on domestic implements, then it is easy to confuse the training that a designer of these objects needs. Is he to be trained to draw realistic roses in order to put them on to firescreens, or is he to be taught how to design good firescreens? Indeed, is it possible to design good firescreens without planting roses on them? So the Victorians' visual sentimentality which tended to blur the dividing line between art and design caused the difference – on a practical level – between fine art and design at the newly established schools to disappear and made it impossible to train industrial designers as such in this country until the Thirties. The good craftsman who could not stand the 'aristocracy' of fine art in the schools disappeared into handicrafts and those who stayed and were supposed to be trained for industrial design, sank in vagueness, irrelevance and impracticality to which the fine arts invariably degenerated when it came to practical industrial design. There had occurred a seepage from the design field into art but there everything became rather nebulous and impractical. (Notes on this theme are to be found in Appendix C.)

Until recently it had been axiomatic that art and design, in the Victorian sense, were closely connected. Thus the First Report of the National Advisory Council on Art Education, published in 1960, accepted the seminal importance of the fine arts. To quote:

> Fine art teaching in a school – whatever the form fine art takes – can serve and we believe it should, as a focal point to strengthen aspiration for the whole school. It is through this teaching (of fine arts) that students will learn something of those mental skills and disciplines which underline and sustain any form of specialisation in art and design.

Only in 1972 did the Report of the NACAE, a joint Committee on the Structure of Art and Design Education, say:

> The study of fine art is not now regarded as necessarily central to all diploma studies in the design field.

This joint committee report suggests three kinds of art and design education:

1. A three-year course at university level leading to a general Diploma in Art and Design – DipAD (after a one-year foundation course) now in many schools, a BADes.
2. Specialised design courses, also at university level, lasting four years and including a period of three to twelve months in industry which is attested by a different Diploma; now, in some

schools, a MADes.

3. A Design Technician's course of two or three years' duration.

Little is said about the training of teachers as such. Teachers were, in the past, supposed to obtain their DipAD and then receive a year's training at a teacher training college. Today they are expected to get their equivalent degrees and then go on to a teacher training college.

The changes which have taken place in the last twenty-five years in regard to the training of artists, designers and teachers have been desirable but could hardly be said to be decisive in character. The entry qualifications for the courses have been lifted a little and now a number of O levels or a smaller number of A levels are normal requirements. The introduction of the Foundation Course and the extension of the Higher Degree Course by one year can, to some extent at any rate, be considered as definite steps forward. The recognition of art courses by academic awards is also – in a purely formal sense – an advance.

If we ask ourselves how does the actual training of the artist, designer and art teacher differ today from the training given twenty-five years ago, we must answer that it differs but not very much in essence, for there is more of the same but nothing new: slightly higher entry qualifications, a longer stay at art college and higher sounding academic awards at the end, all of which are very good and desirable because these mean a more rounded designer and artist, and a better educated one; also possibly greater starting competence; but basically the training, its aims and the qualifications haven't changed much. The kind of training art teachers receive and its relation to modern living and working conditions leaves out much that seems important

It is also evident from the examination of the facts that, for reasons of economy, artists, designers and teachers receive basically the same training and that such differences as exist are only peripheral. One would have thought that this was basically wrong since the tasks of the artist, the designer and the teacher are so very different: the artist has to see, to understand and to react in visual forms, to be the conscience and the mouthpiece of his generation. The designer has to solve practical problems for everyday living, concerned with two-dimensional or three-dimensional objects in such a way that not only the tangible questions are resolved but that intangible, unspoken, human feelings and considerations are satisfactorily accounted for; the teacher has to capture the imagination of the pupils by being able to convey, in exciting terms, the personal relevance to each pupil of the artist's and the designer's work. He has to encourage pupils to experiment and find out for themselves and he has to teach them to form their own reasoned opinions. Surely all these different ends have to be achieved by different means; because the student who aims at becoming an artist is very different from the student who wants to teach, their training should reflect this difference.

Lastly, we come back again to one of the themes of this work: in spite of evidence to the contrary we persist in believing that what art teachers have to teach all their pupils is a degree of creativity and that this, in itself, will make them, eventually, into discriminating users. It seems to me that this is not so, and that to correct the evident and the hidden shortcomings of our system we should change it by recognising and implementing the new approach which is required for the introduction of a new kind of teacher who would teach a new subject.

A university and art-school education based on visual insight, sensitivity and an up-to-date knowledge of current industrial production could prepare young people for teaching in a way which is now largely neglected. Then the proper disciplines of art knowledge and design appreciation could provide the framework for a general educational system which does not completely sacrifice the senses to the intellect. These new-style graduates would not expect to become professional artists or designers; they would look forward to their work as teachers of new subjects, as managers and as organisers, as entrepreneurs and as civil servants. They might well become constructive critics in papers, magazines and television and in lectures, engaging their talents and understanding in the explanation of the work of artists, designers and industrial products, bringing insight and sensitivity to focus on problems of the human environment.

Efforts should also be made to help these new-type graduates to be certain of their choice of career, to clear up and to establish for themselves that what they think of as their talents are indeed mature and genuine qualities and not the results of auto-suggestion or of fortuitous pressures and circumstances, so that those who would make good teachers of this new kind are not trained to become creative artists and designers and that those whose strength is in original thinking, individual expression or in problem-solving are not made to attend courses which are ideal for teachers, managers and critics but are beside the point to people with their special qualities.

All students and especially art-school students also need as wide an education as circumstances permit. Narrow-minded teachers without broad sympathies, without a large fund of tolerance, without the willingness to continue learning and the readiness to examine objectively other views than their own are useless at best and an active menace to their pupils most of the time.

Illiterate artists and designers, encouraged from an early stage in their lives to self-indulgence, to disregard without proper consideration the views and the conventions by which the world lives, cannot, by definition, be of much use to this same world with which they are profoundly and mindlessly out of sympathy. Designers with whom it is difficult to communicate, who pay scant attention to briefs, who lack self-criticism – all by-products of a wider education than most of our designers receive – cannot be of much use to the community in which they live or indeed to themselves. A designer without profound curiosity about many of the aspects of the world in which he lives cannot, by the nature of things, be a good designer for the everyday things that this same world needs. Wide interests, the ability to understand the language of other disciplines and discipline in itself are some of the results of a thorough, non-specialist, broad education.

Most democratic, industrial and super-industrial societies have tried to overcome the handicaps of the time lag involved in all educational investments – the time which has to elapse between the recognition of new needs, ideological sowing and practical, tangible reaping – by the introduction of adult educational programmes. These are intended to make up for the lost learning opportunities of earlier generations and can function on a purely personal level – moulding the adult to complete the rounding-off process which did not take place at an earlier stage in life – or on a broader level by the dissemination of specialised information to producers and to users so as to improve the quality of a wide variety of products at the point of design and production, to improve the value and the services which are part of the product (and so, indirectly, the sales of products) and incidentally to improve the quality of life.

In Britain the rôle of the short-term and intermediate-term catalyst of industrial thinking and improver of industrial design and products has been allotted to the Design Council. It was also originally allocated a certain rôle in the educational system. The pictures which follow reflect to some extent the manner and matter of some of our schools of art and crafts. In the next chapter I shall try to look at the work of the Design Council and at the practical results of its efforts in recent years.

XIIIa THE OUTPUT OF OUR TOP SCHOOLS OF ART

The pictures which follow show the Diploma or post-graduate work of students from some of the top Schools of Art and Crafts in Britain. In an article published in conjunction with the opening pictures of this Section, Lord Esher, Rector of the Royal College of Art, said: 'The story of art education in this country is a success story. I am pretty sure it is the best in the world.' (*Vogue* magazine, September 1974.)

Philip Cook, tea service, RCA 1974.

Diana Gill, modelled fruit teapot, Central School of Art & Design 1974.

Peter Welch, earthenware ground laid coffee pots, Central School of Art & Design 1974.

Philip Cook, cup and saucer for Tea Council, RCA 1974.

Julia Bullmore, decorative embroidered wall panels, RCA 1973. (*Design magazine*)

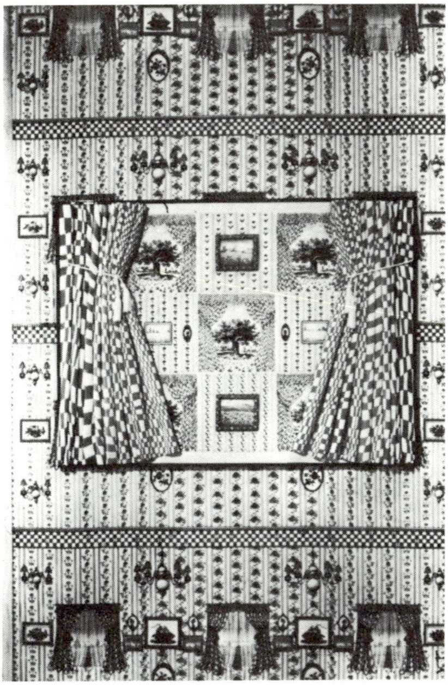

June Bradbury, furnishing textile design illustrating traditional printing techniques, RCA 1973. (*Design magazine*)

Joanna Buxton, woven jacquard textile, RCA 1973. (*Design magazine*)

Janet Bonehill, furnishing textile design, RCA 1973. (*Design magazine*)

Students of Cordwainers, Leicester Polytechnic, RCA and St Martin's, 'Footwear' sketchbook, 1974. (*Reproduced by kind permission of SIAD's journal Designer*)

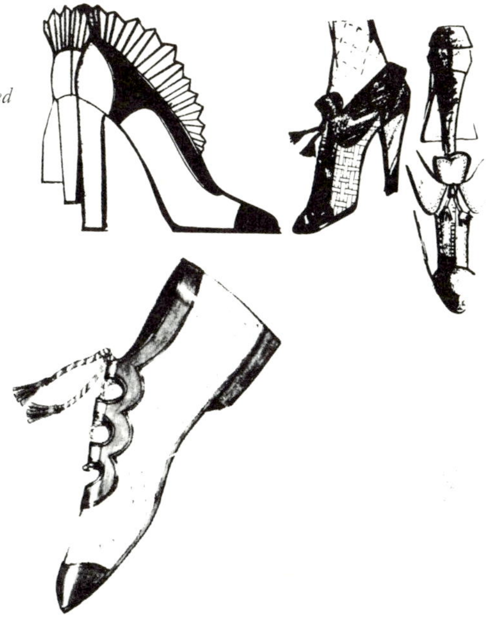

Joanna Bowring, three-dimensional figurative woven tapestries, RCA 1973. (*Design magazine*)

Rosemary Purdy, experimental designs featuring beads in woven textiles, RCA 1973. (*Design magazine*)

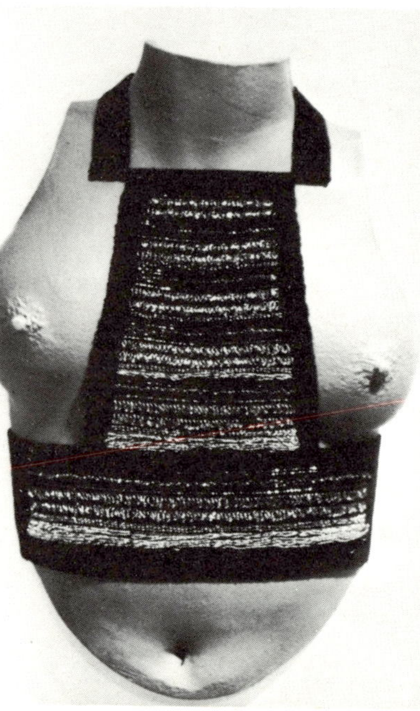

XIV

In a cogently and, to my mind, brilliantly argued series of articles published in April 1973 by *The Economist*, Norman Macrae, its Deputy Editor, diagnosed the British disease as stemming from some of the positive rather than the negative qualities of the people at the top of the institutions of this country. According to Macrae, because of these people and their character, almost all of the changes which are needed are the opposite of all the changes that are likely to be put into effect. In his opinion the great institutions that Britain has created – Parliament, the Civil Service, Local Government, the Law Enforcement Authorities and other public services, the Trade Unions – are not performing the functions which were their original purpose. Most are being operated not to cater for the real needs of the country but mainly to allow the able and upright top people in each of them to enjoy themselves by doing the things that give them most satisfaction and which they themselves do very efficiently. Most of the changes which are needed would cause maximum disruption in what most public men in the country have come to regard as their sense of purpose. These men will therefore have honestly persuaded themselves to oppose such acutely necessary changes.

Macrae's thesis is that the world as a whole has done flabbergastingly well in the years since the war but Britain has done surprisingly badly. He argues that Britain is in a deteriorating situation despite the intermittent application of all the conventional remedies by Britain's above-average governments and bureaucracy.

In Macrae's words, 'The secret canker that is gnawing at Britain's vitals is that we have devised or inherited, throughout our Government and business, a generally brilliant administrative machinery for picking and promoting people; one that may lead the world in elevating the best immediately available men to run the existing system.' These people make the existing system work very well indeed but, he adds, 'At a time when the system itself needs to be changed this is quite disastrous.'

My own thesis in regard to the Design Council is that everything that has been said about our institutions in general is true of the Design Council in particular. It is run by the same kind of people and broadly on the same principles as other Government departments and it suffers from the same virtues as many of our other respected institutions do; but it also has some specially damaging virtues of its own.

The Design Council is not a Civil Service department or a privately subsidised foundation; and it is not a business enterprise. It is a compromise of compromises, and it shows. Resembling the British Broadcasting Corporation in some ways (officially independent but indirectly Government-controlled), staunchly presenting only the British point of view instead of comparing British products with what other people are offering to the world, it is obviously a very different kind of animal: controversy and a critical approach are inimical to it. Officially conceived to increase to healthy proportions the aesthetic appetite of the British public and correspondingly to strengthen the aesthetic muscle of British industry, it has, over the years, evolved an esoteric game of 'musical chairs/spot the trend/a thousand and one exhibitions and conferences', which are marvels in their own spheres but seem to me irrelevant to our industrial and economic problems. If the Design Council disappeared tomorrow, I think British industry would not bat an eyelid and the public would not notice its demise.

The Design Council has clearly and for many years now been part of the establishment. Its Director is answerable directly to Parliament. It enjoys Royal encouragement. It recruits onto its Committees those who have 'made it' and who are either accepted or connected professionally or socially in various fields related to design. The Design Council has, for decades, served as the fashionable Arrivals Hall rather than the workaday Departure Lounge for original and other designers.

The system is that the Director, whose appointment is permanent, influences the appointment of the members of the various committees through which the Council works. This would seem to mean that one man, usually past middle age, controls Council policy and execution for anything from ten to fifteen years. I question the desirability of this system in any administrative or industrial context. In an area where change, adaptability and flexibility are more important than in any other sphere, one can only think that this system is bound to have grave disadvantages.

One of the worst things that can be said about the Design Council is that it is thoroughly cliché-ridden; the best, that its clichés are bang up-to-date. Another bad thing that could be said of it is that it falls constantly between an infinite number of stools; but it is perhaps just as well that they are, every one of them, stools of great contemporaneity designed by 'the best' people. One could also say of it that its bite must be worse than its bark (for it almost never barks); but then, that it has no time to bite; it is too busy chasing its own tail.

A brief summary of the Council's annual report for 1972/73 is interesting at this point in the discussion of its activities. (For a more detailed summary see Appendix F.) The Design Council, as portrayed by itself in this report, emerges as an efficiently run, busy organisation with devoted, able and competent, not to say hard-working, people at the top and on the staff who obviously believe in and enjoy what they are doing. But are these the things, one is compelled to ask, that really need doing?

In 1972/73 various committees, sub-committees and advisory committees met twenty-three times (not counting the meetings of the Scottish Committee). Five meetings were held with other organisations. One tour of industry was organised. Services to industry were attempted involving Council staff in twenty-one visits. Forty-six Awards were made. Thirty-two conferences, courses and seminars were arranged. Three Scholarships were awarded. Two Educational Trusts were administered and memoranda on educational matters were written and submitted. Three hundred and fifty-six enquiries for designers were dealt with (no record of subsequent actual engagements is shown).

Six hundred and seventeen groups, comprising nine thousand one hundred and seventeen people, visited the Design Centre and, of these, two hundred and sixty-one secondary school groups and ninety-four other groups were given short introductory talks. One million, seven hundred and sixty-six thousand, four hundred and fifty-six people visited the Design Centre at the Haymarket to look at fifty-three special displays out of which eight were thematic exhibitions. There were six exhibitions abroad. The Design index continued to list nearly ten thousand items and about thirteen and a half million Design Centre labels were sold. The circulation of the magazine *Design* at seventeen thousand, seven hundred and ninety-six was a little lower than the year before. Two new Design Centre books were published and a number of others reprinted. Two retail-selling booklets, a synopsis of a conference and a directory of design courses in the United Kingdom, were also published. Twenty-eight press conferences were organised and four thousand five hundred and seventeen mentions were achieved. Radio and television coverage of the Council's activities were reported to be very satisfactory.

How does this abstract but factual summary of the Council and its work – and the impression which it creates – line up with the realities of industry and life in Britain?

At the end of March 1974 an important but industrially barely noticed event took place in London: the long awaited *Guide Michelin* to Great Britain and Ireland was published and launched, the first in more than forty years. The significance of the appearance of this book was not so much in the quality of the writing or in the quantity of new ideas expressed in it, but in the fact that, with the help of the *Guide Michelin*, it was possible, for the first time, to draw true parallels, in a given industry, between Britain and another European country, in this instance, France.

In the opinion of the *Guide* not one restaurant in Britain or Ireland was worth a special visit or detour. Only twenty-five restaurants were awarded the coveted rosette for good food, nine of them in London, two in Ireland, one in Scotland and the rest in England and Wales.

France has five hundred and forty-nine restaurants in the same category as the twenty-five British ones which were awarded the rosette, not counting an additional seventy-five which are altogether in a higher category.

The care and time taken by the experts to compile the *Guide Michelin* for France are well known; the same care and the same amount of time has gone into this compilation and, what is more, the inspecting staff was the same in Britain as it has always been in France. There are not many industries where direct comparisons between countries are possible; the restaurant industry, however, is now one of them and here the *Guide Michelin* merely confirms what most people with experience have thought or known for a long time.

It is felt by many that, were a similar comparison made in regard to British industry in general and its counterparts in North Western Europe and North America, in relation to design as a whole, the conclusions would be not unlike those reached by the *Guide Michelin*.

Fortunately for the British catering industry, it does not have to export its indifferent food. It does not, by and large, have to compete with French and other restaurants; being situated in Britain it can comfortably exploit a captive audience who either eat what they are offered – to the accompaniment of some expressive shoulder shrugs – or starve. Unfortunately for the rest of British industry it does have to compare its own products with the products of Continental manufacturers which are more acceptable to Europeans and to Americans.

Even more unfortunately for British industry there is nobody to perform for it the function which the *Guide Michelin* performs for the catering industry. It seems to me that instead of a comparative and nationally unbiased approach to its products, which British industry should welcome, it receives views and comment from the Design Council which is often subjective and in favour of anything British, simply because the Council believes that its function is the promotion of British products.

If our restaurants do not come out well from the *Guide Michelin*'s comparisons, it would be wrong to put the responsibility for the quality of the services of the British catering industry entirely on the industry itself. There are a number of other factors involved, the most important amongst which is the quality of demand in this country. It is realised well enough that, were one of the five hundred and forty-nine rosetted French restaurants (not to mention one of the seventy-five super-rosetted ones) to be moved to, say, Birmingham, Bristol or Blackpool, it could not survive for long even though they all seem to be thriving in France. The reason for this is quite simple – there is a demand for that kind of food in France, just as clearly as there is not a demand for it in Birmingham, Bristol or Blackpool. On the other hand one does not have to be a prophet to predict that, if a profitable smoothly running Blackpool restaurant opened a replica of itself in, say, Bordeaux, it would be out of business very quickly indeed for there is no demand for its kind of food and services in Bordeaux (even if the prices are lower – which itself does not necessarily follow).

Because this situation, in general terms and in regard to comparisons between certain British industries and their competitors abroad, is not new and because it was recognised as a concrete fact long ago, the Council of Industrial Design and the organisation which preceded it were formed in order to try and do something practical about design which is essentially immeasurable, elusive, intangible and abstract. The rôle of the Design and Industries Association, of the Council of Industrial Design and of the Design Council has always been and, as far as is known, remains the promotion of a good-design orientated demand by the public and the supply of good-design orientated products by industry.

The years which preceded the foundation of the Council of Industrial Design were formative and important for our attitudes to design today. At their best they are characterised for the man in the street by the work of Sir Herbert Read and a few others, who made it clearer than it had been before then that design is not something that can be added to an existing product, but that, if it is to be valid, it must be part of the product; that a designer is not someone who is called in to prettify things but one who is concerned with the basic concept of a product and with its making. It became accepted in the Thirties that, in a work containing an element of design, as in art, there must exist an irreducible kernel of originality and honesty at the centre from which the work can grow outwards; and that the reverse, a growth which is imposed on a product from outside, cannot result in valid, good design.

The design-orientated thinkers of the Thirties were also quite practical. They saw that to implement their kind of approach to industry, two preconditions had to apply: there had to be an enlightened public to demand well designed products and, no less important, to reject badly designed ones; and there had to be enlightened manufacturers, not only to meet this new kind of demand, but who would sometimes try and stretch the quality of demand into new and forward-looking directions. Since it was never quite decided which comes first, enlightened manufacturers or an enlightened public, and since it was fairly clear that manufacturers considered it their primary function to supply the existing demand rather than to educate the public into new and an aesthetically better kind of demand, it was agreed that the rôle of the Council of Industrial Design was to work closely with both the public and the manufacturers and to make this new and recently established view of design widely known and accepted both by the public and by industry.

One of the most immediately striking things about the Design Council today, therefore, is that such a major part of its activities in recent years seems to be concerned with preaching to the converted, or worse – aimed at trying to achieve instant conversion – rather than with steady and patient explanation and education; with the happy assumption that it is quite possible to convert a philistine manufacturer into a seeker of true and good design at one of its conferences or during one of its exhibitions; that, while admitting that good design cannot be smeared on to a product, it is possible to maintain that good intentions, good resolutions and lots of knowledge about good design can be so smeared on an unprepared and, in design terms, quite unaware human being.

Thus, among the Council's main activities there was the planning and the organising of conferences for industry; the annual report enumerates close on thirty such conferences, courses and seminars attended by a total of about fifteen hundred people and seems to assume that all these people crowded into the conferences, thirsting for the truth about design and eager to be converted to the Council's point of view. Even if the number of those attending these conferences was significant in an industrial context, which is doubtful, one might be forgiven for suggesting that it borders on the naïve to assume that companies and institutions do not contain many people who are only too glad to get away from routine to any conference; that it is optimistic to expect more than a few papers at such conferences to contribute much worthwhile new information; and that it is simple-minded to pretend that conferences of the kind arranged by the Design Council make any real or lasting difference to industrial knowledge, to human relations, to quality, to production or to design in Britain. To believe that they actually convert the ignorant and the cynic in aesthetics and design into believers in visual goodness is no different from thinking that Mr Billy Graham, the Evangelist, actually converts, in one moment at one of his well-publicised meetings, religious non-believers into permanently devout and reasoning Christians.

I find it difficult not to believe as a general rule of life that those who can, do; that those who do not want to do, teach; and that all the others, who can neither do nor teach, attend conferences.

I am unimpressed by a long list of, and I dislike, conferences, even though I have had my share of them, because it seems to me that they have achieved little in the past and seem unlikely to achieve more in the future. One does not learn to do things nor does one actually do them at conferences, but in private. Company executives who want time off or a change would be far better off looking around a foreign city of their choice or playing golf or some other game. They should not allow themselves to become the honestones of other people's egos.

The intention of the Council's founders was, one would imagine, to use it as an instrument of change in matters of design and taste, as a catalyst of new thought, facilitating the acceptance of new ideas. We may therefore be justified in asking ourselves and the Council whether all these conferences really add up to anything substantial and so fulfil the founders' intentions; or is the Council merely unable to stop doing what it started doing many years ago, regardless of the fact that these activities are less relevant nowadays, and should perhaps have been replaced with newer methods of promoting the Council's objectives?

Then take the committee meetings of which around thirty are mentioned (including the meetings of the Scottish Committee). An impressive total, one might say, representing roughly three hundred working days by some of the country's top design experts (if one assumes that ten design experts attend each meeting). Surely something worthwhile must happen as a result of such a concentrated effort. The facts, to me, seem to be different.

I myself must have attended something like fifty such meetings in my years on the Scottish Committee of the Council of Industrial Design (1964–72) and I can say with some conviction that I cannot remember many occasions on which much of importance to industry or to the country was said or done. It is, of course, true that I have a bad memory. Worse, I have a highly selective memory. So, apart from not being able to remember anything positive, I find myself stuck with some rather negative recollections: like the occasion on which the Scottish Committee as a whole found itself at such a distance from the conclusions of the London Design Awards Committee that it thought that the award-winning designs were unworthy of any awards. I remember, too, the wry taste of politics – nationalistic, chauvinistic, parochial – which some of the discussion of the Scottish Committee left in my mouth and the vague feeling of being fobbed off 'by London' which many of the discussions left in my mind; I remember the sadness and embarrassment at sensing the Kafkaesque conditions under which some of the Council's permanent staff seemed to work.

How can one forget the feeling of total frustration which was experienced almost each time one returned home from one of the meetings which, with travelling, usually took up the whole of one day. Another day wasted. Why not resign? Because perhaps next time one will be able to do something worthwhile, or the next time, or the time after that. The thirty or so committee meetings mentioned in the report leave me therefore absolutely cold.

The report mentions Services to Industry. It would be surprising if an organisation which has on its staff some of the nicest people one could wish to meet – but which also, by its very nature, must contain the least professionally creative members from a number of professions and the least industrially experienced despite their other highly desirable qualifications – could, through its staff, offer any real constructive help to industries and to companies which owe their existence and positions to the fact that they employ the most industrially experienced and professionally active people. There is very little indeed in the way of specific services – apart from publicity and the display of products at the Design Centre or elsewhere and at certain exhibitions abroad, that the Council has ever offered on a practical level in the past or can offer now or in the foreseeable future.

Trade and industry have always had reservations about the Design Council. There have been quiet, joking and no doubt exaggerated asides, both by those whose products were approved and indexed by the Council and by those who could not get their products into the index, saying in fact that such approval seemed to them – in commercial terms – the Kiss of Death. The Design Council's awards themselves are also jestingly said by some to be a Fate Worse Than Death. The thirteen million Design Council labels sold over the past year seem to be telling a different tale, one realises that; but then perhaps it is simply that trade is superstitious. The Design Council labels are not the only talismans with which merchandise is bedecked and hung so as to encourage sales. Like the ancient Romans, some businesses may sacrifice even to Gods of whose standing and favours they are not completely certain but whose help they would like to enlist.

I have said earlier that, in design as in anything else, judgments should not be too personal and subjective and that they should never be final and absolute. Even though full objectivity can never be achieved, for subjectivity to be acceptable it requires certain minimum conditions and qualifications. The awards which the Council makes are, I think, meaningless most of the time and quite harmful some of the time because the main considerations in reaching decisions seem, to my mind, to be those that should not matter. I must emphasise that I do not suggest for a moment that there is any kind of corruption or dishonesty at any time. On the contrary, the interests of the Council are always uppermost in the minds of those who decide the award-winners. I am, however, saying that the interests of the Council and of good design, which are not always identical, are misunderstood, misinterpreted and misjudged from the highest possible motives. Like every executive body, the Council needs good administration and good clubmen; but it needs just as much, if not more, good designers and thinkers who are fairly articulate, are not intimidated and speak their minds.

It is said in the report that three hundred and fifty-six enquiries were dealt with in the course of the year under review; but it is not stated how many of these actually resulted in commissions, which, for an organisation interested in the promotion of the practical aspects of design, is strange. However this is a really important and worthwhile activity; one which, in any well run business, would account for the time of a design-orientated administrator and perhaps a part-time assistant. The puzzling aspect of this 'service' to industry is why industry or commerce should not, in fact, deal with such enquiries better (and without charging a fee) than the Design Council does, or not be charged a fee unless a commission actually materialises. This already happens commercially in fashion design and it even happens in art where the galleries to whom artists are under contract (or even when they are not) will recommend them for certain commissions and help with negotiations and administrative details and charge nothing if no commission results. As far as one knows, the Architectural Association does not charge a fee for recommending architects for large or small commissions.

Six hundred and seventeen groups are said to have visited the Design Centre and, out of these, two hundred and sixty-one secondary school groups and ninety-four others were given short introductory talks. This means that something like five thousand school children visited the Design Centre over the year. The remaining ten million children who did not visit the Design Centre and whose teachers have barely heard of the Council's activities (see Appendix C), are not mentioned in the report. It would be interesting to know exactly what the Council is doing in regard to helping to promote design education in primary and secondary schools and at universities. One would have thought that this kind of work would be one of the major projects in the Council's programme and that it would be mentioned in the annual report.

The figure of one million, seven hundred and sixty-six thousand, four hundred and fifty-six visitors to the Design Centre at the Haymarket is given in the report and one cannot but be impressed at the thought of all these earnest seekers of good design solemnly trooping through the Design Centre's doors, being carefully counted and recorded, studying exhibit after exhibit, taking notes and making comparisons, valuations, decisions and generally behaving in an enlightened, responsible and progressive manner. Until one remembers, that is, that the Design Centre is not that far away from Piccadilly Circus and could be so convenient as a rendezvous. It is dry inside and there is no charge and one can look around while one waits. It is just possible that it may be relevant to compare this staggering figure with the even more staggering one of the number of people who pass through Piccadilly Circus yearly or with the figures of the customers of one of the shops in the neighbourhood with a similar floor space to the Design Centre. Or to put it in a different way: fewer people in a year have looked at the designs in the Design Centre than read, no matter how quickly or carelessly, the *Daily Express* or the *Mirror* or the *Sun* or the *Daily Telegraph* on any one day – or the *Sunday Times* on any single Sunday of the year. Do these attendance figures mean anything therefore in terms of persuasion and the dissemination of new ideas or is this simply an all-year-round jamboree for converts and 'waiters'? But, if they do, what a marvellous chance for disseminating the idea of good design is missed, for the Centre shows little which is substantially different from the merchandise that can be seen in most shops.

Talking of newspapers, it might be as well to consider here the figure of four thousand, five hundred and seventeen mentions in the Press. The truth is that the Design Council has a pretty poor

Press on the whole. It is important that it should be realised that the Design Council mentions in the Press are in no way mentions of the Council's ideas, aims and achievements: rather they are similar, with some exceptions, to the mentions of, say, Olympia as the venue for events from Cruft's to Camping Exhibitions. Hundreds of firms and organisations are persuaded yearly to show in the Design Centre: they pay the Design Council for their space and the other services that go with it and the Press writes, if they write at all, about these firms, organisations and their products, mentioning naturally that they are being shown at the Design Centre. It would be fair to attribute the publicity gained from such exhibitions not mainly to the efficient deployment of the Council's budgetary and other resources but to the efforts of the public relations officers employed by the firms themselves and by the organisations involved in the exhibitions. Design Council events as such, even the announcements of or the presentations of Design and other Awards by the Duke of Edinburgh receive comparatively scant comment in the Press.

Lastly we come to the consideration of some of the policies which the Design Council has adopted in recent years. As its new title implies, there are some fundamental differences between the old Council of Industrial Design and the new Design Council.

Engineering, it has been argued, is responsible for by far the largest proportion in the value of exports from Britain. We must therefore, the argument runs, make certain that all engineering products, from earth-moving equipment to filters for sludge-extractors, are well designed. The accent of the Council's activities must consequently be shifted from consumer goods to industrial equipment and engineering products. As the introduction to the 1972/73 Report states:

> The more the Council looked at its total field of activity, covering as it does engineering, industrial and graphic design alike, the more convinced it became that the Government had taken the right decision* in inviting the Council to pursue its campaigns beyond the boundaries of industrial and graphic design even more vigorously than before into the vastly more powerful and complex realms of engineering, for in engineering the disparity between a designer's responsibility and his authority is sometimes more marked than in other industries.

It was of course frustrating not to be supplied immediately with

* That is, not without some lobbying and persuasion

funds to expand the Council's activities. Nevertheless, by careful management of available resources it was found possible in September to appoint a Head of Engineering Design together with a small nucleus of staff. With the support of the eminent members of the Council's Engineering Design Advisory Committee, good progress was made in preparing the ground against the day when an increased grant-in-aid would enable a real impact to be made.

The Council was well aware that without a considerable increase in qualified staff only the surface of the task of promoting British engineering design could be scratched. However these preliminary canters clearly revealed both the need for and the range of opportunity awaiting a properly financed initiative. The Council was therefore confident that its thoughts would be well taken within the Department of Trade and Industry and that adequate funds would be forthcoming.

Although it was natural that the Design Council should, in the first year of its new title, spend a great deal of thought and time on the new activities that would justify the change of name,* the areas of design associated with the former Council of Industrial Design were by no means neglected, since it is a cardinal belief of the Council that design is indivisible and that no aspect of the total creative activity should be promoted to the exclusion of or at the expense of any other.

For an organisation which proclaims the Indivisibility of Design to be its cardinal belief, this decision of launching itself into engineering design is inexplicable because, translated into practical terms, it means that the Council expects an aesthetically badly housed, badly furnished and badly clothed people, in fact, the inhabitants of our brand new industrial slums, who are apparently without much feeling for the man-made or the natural environment, to care deeply about the shape and the colour and the proportions of their sludge-extractors, earth-diggers and cranes. One can imagine letters appearing in the *Daily Mirror*, written by council tenants protesting about the unpleasant shape and colour of the earth-diggers working in their vicinity on extending the estate. Even allowing for the British special relationship to the car (how else explain those glowing, lovingly looked after and comparatively expensive cars in front of dull, dingy, ordinary and comparatively inexpensive flats and houses), this stretches credibility too far. It would be quite natural for people who do care about the

*This, in parenthesis, seems a remarkable proposition in itself: fundamental policy decisions are, we appear to be told, being taken partly in order to justify the Council's new name. Was there ever such a tail which wagged such a dog?

appearance and the aesthetic content of the little, daily things in life to care also about big things which they produce or use; for those who are prepared to spend their own time and money in order to achieve practical and aesthetic satisfaction in the objects which they make or buy and with which they surround themselves, also to do their best to achieve a similar relationship in regard to engineering and capital goods which they do not use daily but with which they may be connected in one way or another or which they see only on special occasions. Indeed this would be a natural way in which one would expect people to behave; but the opposite cannot be true. People who are visually slovenly or uncaring in little matters will be equally sloppy and insensitive in important, major projects. One may even go further and say that the way to make people care about the appearance of large projects and infrequently used objects is to teach them early on to care about the appearance and design of everyday things. The grass roots of a country's design are to be found in individual houses and homes. If the contents of the average home are aesthetically sound, the country as a whole and its industrial products will reflect this unmistakably; just as the opposite will be equally truly reflected.

The state of our exports, the appearance of our housing estates, the visual conditions prevailing in factories in our cities, in our public utilities and in our environment in general and, most important, the atmosphere of some confusion in our primary and secondary schools in regard to the teaching of art and the lack of realisation of the alternatives or of the need for instruction in visual awareness, should make it impossible for anybody, even for the Design Council, to argue that having truly laid the foundations we are now ready to go on with the erection of the super-structure and that we can therefore now afford to concentrate on the things which are not of general concern but only the concern of the specialists. It is impossible to maintain that, having educated the public in general at primary- and secondary-school levels and later, in matters of discrimination and taste in regard to everyday things, we are now ready to devote expense and attention to engineering and capital goods.

The concept and the rôle of the Design Council remains, one would have thought, quite straightforward and simple: to create a higher standard of demand from the public and to encourage a higher standard of supply by the manufacturers. To help with the education of the preponderant majority of people in this country to higher standards of aesthetics than those which now prevail and to persuade industry not only to cater for this new need but also to lead, by creating better standards of demand. Experience over the years should have taught us that one of the most practical and effective ways of doing this is at school level where the largely misunderstood concept of 'creative art' through teaching should have been modified long ago and supplemented by a down-to-earth approach to aesthetic awareness particularly in regard to our living and working environment and to objects in everyday use. Had this happened, that is, had we met this self-imposed task over the past fifteen years, it might well have been unnecessary now to plan a concentrated attack on the appearance and the design of engineering products. For then the improvement in the appearance of these products would have happened spontaneously, as a natural sequence to the improvement in the appearance of products used in everyday living. To neglect the foundations – as I believe the Design Council is neglecting them now by not taking its rôle in education nearly seriously enough – and to concentrate effort on the appearance of the superstructure – as I believe the Design Council is doing now by diverting resources from its basic functions to subsidiary functions – is wrong and futile. The Design Council has never, as far as one can see, played to the full its intended rôle of being a really important guiding and helping factor in education on a national level.

Among the virtues of the Design Council, one could consider the fact that, in theory at least, it is an effective tool in publicising good design because of the qualifications of its executives, some of whom are journalists by profession and are naturally publicity- rather than design-conscious. This, however, has perhaps had an unhappy effect on its relations with the Press who probably feel that they can do little for an organisation with many strong-minded people whose pronouncements on design carry little more weight than their own do on this subject. The following extracts from *The Times* diary of 15 August 1975 reflect not only the prevalent atmosphere at the Design Centre and certain of its attitudes to design, but also gives a fair picture of its relations to the press:

Peter Shore, the Secretary of State for Industry, was promoting his argument that people should buy British goods whenever possible by touring the Design Centre in the Haymarket and delivering a pep talk at a press conference afterwards.

[He] took his walk round the Design Centre at a pace too rapid for the crowd of press photographers his visit had attracted. They complained that Shore would not stop and look at anything interesting.

'The colleges chose these. We didn't,' said Sir Paul Reilly, director of the Design Council, apologetically when the party reached a section devoted to student designs. The photographers clamoured for the minister to examine a chest of drawers made like a green hippopotamus but Reilly refused to allow it. 'It comes as a cocktail cabinet, too,' murmured one of the officials walking behind.

At the press conference the photographers trained a wide variety of Japanese, Swedish and German cameras on him and radio reporters recorded his words on Swiss and Dutch tape-recorders.

Reilly presented Shore with a Union Jack table-cloth and a pair of socks in Union Jack design – souvenirs, he said, bought from the Design Centre's own shop.

One could say, as the Design Council probably does, that these things happen and that one cannot expect better relations with the media in an organisation like the Design Council. One would, however, have expected the Council's own magazine, which is totally and directly under the control of the Design Council, to be an example of what a magazine, and a design magazine at that, should be. An industrialised country which boasts of some of the best daily newspapers in the world, has undoubtedly the best weeklies and monthlies on politics, sociology, the arts and science, has every reason to expect a good monthly on design. Such expectations lead to disappointment as far as *Design* is concerned: in my opinion, sad while trying to be snappy and irritating instead of entertaining.* It would be so easy to be stimulating, controversial and, on occasions, even scintillating when writing about and illustrating design. Yet the adjective one hears most often used to describe *Design* magazine is 'annoying'; for, strangely, it manages to seem arrogant and timid at the same time; ponderous and empty; more than fully illustrated yet I find it lacking illumination and imagination, not only in content but in layout.

One of the great virtues which endlessly handicaps the Council is the very British virtue of compromise. The Council is not a

*These remarks, although applicable to *Design* magazine over a long period, may happily no longer apply, since at the present date (June 1975) changes are being made in the magazine.

commercial enterprise, free to comment, condemn and compete; it is mainly Government-subsidised, is run on Civil Service lines and is, by its constitution, inhibited from doing any of these things wholeheartedly and freely; institutional and commercial politics play an important part in its day-to-day activities. Because it has to depend for a substantial part of its income on industry, it has to tread carefully so as not to offend and, if possible, so as to attract the key people in industry.

Ideally, the Council should be financially independent of (but in certain other respects inseparable from) industry and of Government, even though wholly Government-subsidised. Or it should have the courage of its convictions and should be run on purely commercial lines as are a number of idealistic undertakings who rarely mince their words (*The Observer*, *The New Statesman*, *Which* and others). There is no surer way of confusing objectives and then of missing them than by constructing a formal, artificial entity like the Council which will always get the worst of all possible worlds.

The concept of the Design Council is important; its existence essential, perhaps more today than ever before; but its ends and means should be re-examined and, if relevant, brought up to date, clarified and re-affirmed in so far as they still apply to today's conditions: or dropped if they do not apply. The ends should be clearly set out as should be the practical means of their realisation. Some of the points to be considered could conceivably be these:

1. The exact extent to which the Council should be involved in helping elementary and secondary schools with training in visual discrimination in general and with the teaching of the appreciation of the works of designers and craftsmen in particular. If it is agreed that what we now call school training should, in art and crafts, broadly speaking, be consumer-orientated, then it can never be too early to start.
2. The extent to which the Council should be involved in and the help which the Council should give to the planning and the carrying out of the 'new-teacher' training programmes (that is, the training of the new kind of teacher as outlined previously).
3. The more exact definition – in practical terms – of the Council's rôle in adult education programmes, so as to improve the standard of demand even before the new generation comes into its own.

4. A consideration of better ways than those employed at present, of helping industry in design matters.
(a) A better magazine on design: readable, informative on a world-wide basis, with a clear and consistent point of view.
(b) Help with the initiation and the mounting of exhibitions at home and abroad.
(c) The supply of facts on design based sales trends.
(d) Better aids to policy-making at Board level. (An industrial officer is unlikely to influence policy on one of his visits. Men of reputation would.)
5. Better ways of helping designers than at present available:
(a) A genuine advice service by prominent and successful industrialists and designers.
(b) A better job-finding agency.
(c) A mouthpiece for young designers.
(d) Commissions by the Council as exercises in adventurous design in areas where industry fears to tread.
(e) A more demanding and therefore effective selection and promotion of young designers' work.
6. That the Design Council should not be engaged in promoting or in selling merchandise; that is a salesman's job; but that it should promote the idea and the ideology of good design.
7. That the Design Council is too big for the job it is doing at present; that it includes too many people without the most appropriate qualifications.
8. That the standard of recruiting to the permanent staff of the Council should be as high as possible. This means competitive salaries to attract the liveliest, ablest and most ambitious minds from industry.
9. The possibility of remunerative involvement of top designers in the running of the Council's affairs, even if only on a part-time basis and for comparatively short periods of one or two years.
10. That all important executive posts on the Council should be occupied for a period of, say, five years; successful senior people should then be shifted into positions connected with design in public or private ventures so that they could offer these enterprises both their prestige and their experience. That successful middle-management people should move up steadily and be promoted into the top jobs.
11. That a highly qualified and strong board under an equally

well qualified (and paid?) chairman should settle policy matters and see to it that they are implemented; and that the board and not the chairman should propose and elect new members as well as its chairman.

12. The possibility of having the equivalent of a Minister for Design at the head of the Council who would be either elected or appointed for three- to five-year periods.

13. That a clear-cut decision should be reached at Government level, in regard to the aims, status and functioning of the Design Council. If it is decided that it should be a service to industry for exhibitions, conferences and promotions then it should be mainly profit earning; but being the servant of industry it should then not attempt or pretend to be the judge of its products and a factor in the shaping of its aims and ideology.

If its functions are to be educative then it should be completely free from the responsibility of earning profits from industry and from the inhibitions, vis-à-vis industry, which are bound to run parallel with this responsibility. In this case its grants should be increased very considerably (five million pounds would not be too much at today's values) so that it can be totally independent of industry whose design activities it could then keep under constant, informed and public, impartial but friendly, independent and constructive review. This would mean that the Council need show only those products of which it wholeheartedly approved and show them as and when it wished. It would also mean that the Council could spend its money on educational projects which show no direct or immediate returns. In this case the Council's function in regard to industry could be that of a candid and vigorous friend, a factual assessor and comparer of the products of British industry with those of other countries.

To say, as we seem at present to be saying, that the design education of children, of the public and of industry is important in itself and also because it affects thousands of millions pounds' worth of exports annually, that it needs help and the right kind of encouragement but that we cannot afford to spend on it the extra half a million which the Council now earns from its commercial activities does not seem to make even economic sense.

XIVA EXAMPLES OF DESIGN AWARD-WINNING AND OTHER FEATURED PRODUCTS

The following pictures show some of the products from British Industry which have, in recent years, been prominently featured in *Design* and *Designer* magazines, or have received Design Awards. Explicitly or implicitly they have been portrayed as typically good British design.

'Hexagon'

Crossley Carpets. David Hicks design.
Design Awards 1970.

'Crossbridge'

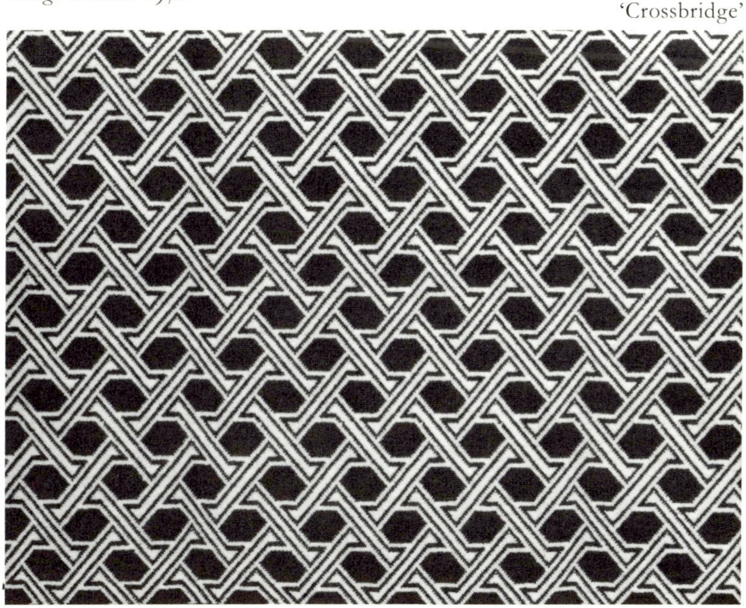

Adjustable mirror,
Peter Cuddon,
Design Award 1961.
(*Design magazine*)

'Strawberry Hill',
Dinner Service, Josiah
Wedgwood & Sons Ltd,
Design Award 1957.
(*Design magazine*)

Dining Chairs, Archie Shine Ltd, *Design Award*, 1963. (*Design magazine*)

Heavy Duty Settee, Hille Ltd, *Design Award*, 1962. (*Design magazine*)

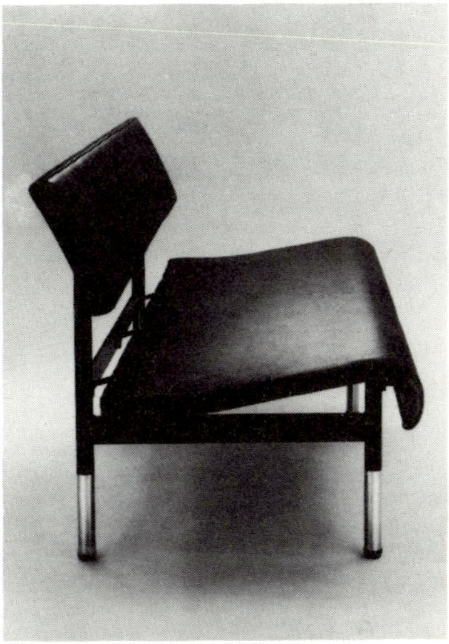

Circular Dining Table, Design Furniture Ltd, *Design Award*, 1959. (*Design magazine*)

'Vynide' Coated Fabric Wall-covering, ICI (Hyde) Ltd, *Design Award*, 1960. (*Design magazine*)

'Royal Gobelin' Axminster Carpet, Tomkinsons Ltd, *Design Award*, 1958. (*Design magazine*)

'Pannus', Wallpaper Manufacturers Ltd, *Design Award*, 1960. (*Design magazine*)

'Kew', Mary Oliver, Heals Fabrics, 1974. (*Design magazine*)

'Amik', Sekers Fabrics, 1974. (*Design magazine*)

'Snowcaps', Julek Heller, Heals Fabrics, 1974. (*Design magazine*)

'Piccolo' bedlinen, Marianne Garsted, Christy, *Living Award for Good Design*, 1974. (*Designer magazine*)

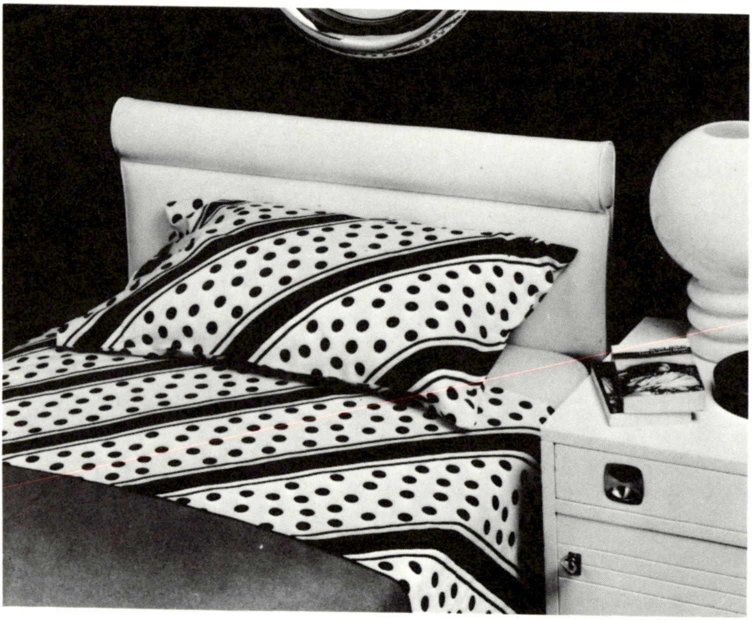

'Herald', Tamesa Fabrics, 1974. (*Design magazine*)

'Sierra', Michael Fieldsend, Textra Fabrics, 1973. (*Design magazine*)

'Automation', Barbara Brown, Heals Fabrics, *Design Award*, 1970. (*Design magazine*)

'Salad Days', Kay Politowicz, Textra Fabrics 1973. (*Design magazine*)

'Chevron', Andrew Foster, Textra Fabrics 1973. (*Design magazine*)

'Paramo', Raye Clements, Textra Fabrics 1973. (*Design magazine*)

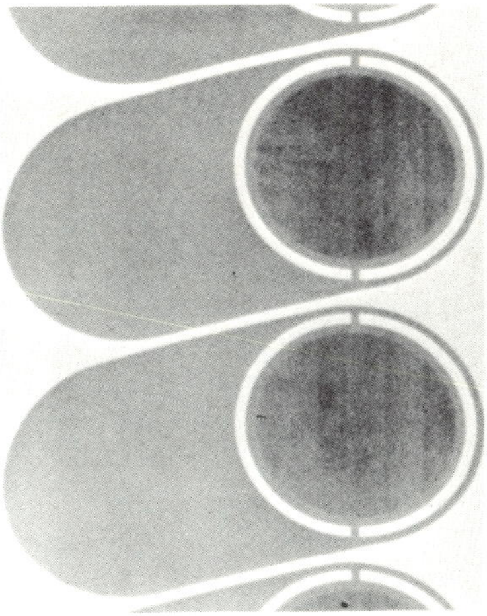

'Meteor', Peter McCulloch, Textra Fabrics 1973. (*Design magazine*)

'Panorama', Raye Clements, Textra Fabrics 1973. (*Design magazine*)

'Minaret' Wallpaper, Osborne & Little Ltd, *Design Award* 1969. (*Design magazine*)

'Red Herring' Wallpaper, Osborne & Little Ltd, *Design Award* 1969. (*Design magazine*)

'Rabat' Wallpaper, Osborne & Little Ltd, *Design Award* 1969. (*Design magazine*)

'Jonas Cord' Wallpaper, Osborne & Little Ltd, *Design Award* 1969. (*Design magazine*)

'Five', Craven, Hull Traders, *Design Award*, 1968. (*Design magazine*)

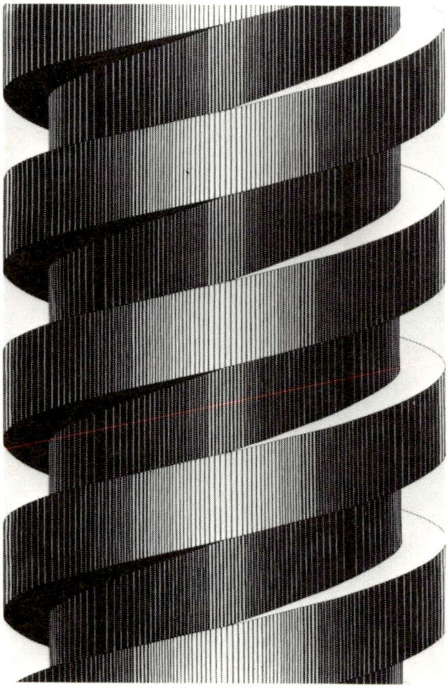

'Spiral', Barbara Brown, Heals Fabrics, *Design Award*, 1969. (*Design magazine*)

'Caprice', vinyl flooring, Armstrong, *Living Award for Good Design*, 1974. (*Designer magazine*)

'Mitre', Catherine Netherwood, Heals Fabrics, *Design Award* 1969. (*Design magazine*)

XIVB SOME EXAMPLES OF GOOD BRITISH DESIGN

My thesis, that we need to reconsider and change some of our ideas about and attitudes to design matters, would be narrow and unrealistic if it did not acknowledge the achievements of British designers.

The following photographs, limited in number mainly by the economics of book production, should make it clear that I have a high regard for good British design.

Spitfire at Biggin Hill Air Display. (*Tony Boxall*)

Supermarine Spitfire designed by Reginald J. Mitchell, 1936. (*C. Bicknell*)

The bath was designed specifically for the shape and safety of a baby. It has a contoured seat which supports the baby in the correct position for bathing and prevents it from slipping under the water. Shaped arm rests in the side of the bath make it easy and comfortable for the mother to hold the baby safely and with confidence. David Harman Powell, Design Research Unit, 1966–7. (*Grey Lacey*)

The lavatory seat is a 'toilet trainer', designed to give maximum security to the young child, boy or girl. The seat, which fits neatly on any standard WC, has a smooth high gloss surface. Being made from compounded polypropylene it is non-absorbent, non-toxic, rigid, sterilizable, resilient and warm to the touch. David Harman Powell, Design Research Unit, 1966–7. (*Grey Lacey*)

Bristol type 411 Saloon design by Dudley Hobbs and his team, 1970.

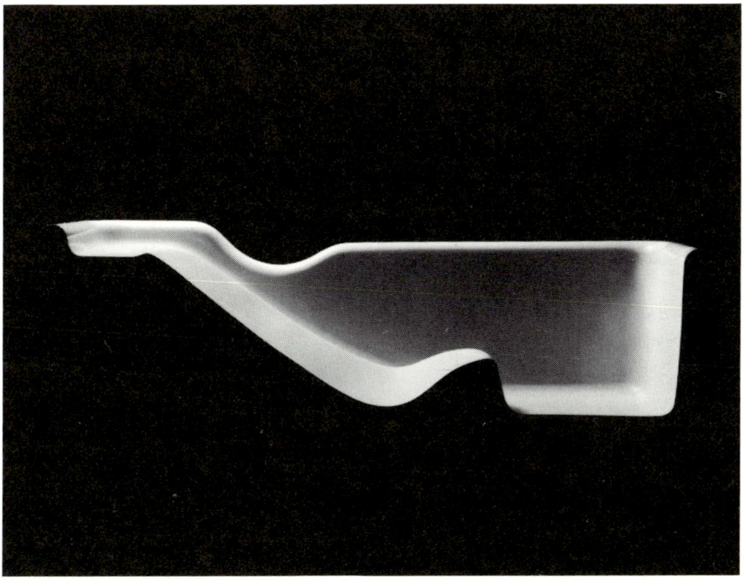

'Quad One' pedestrian lighting bollard, designed by Peter Smart LSIA, 1973, for Abacus Municipal Ltd.

Harlow Town Station, 1960, British Rail Eastern Region. (*British Rail*)

Norgas House, Newcastle – view from Northumbrian Way. Designed by Ryder & Yates & Partners, Newcastle, 1965. (*Philipson Studios*)

Norgas House – restaurant block at night including illuminated sign. (*Philipson Studios*)

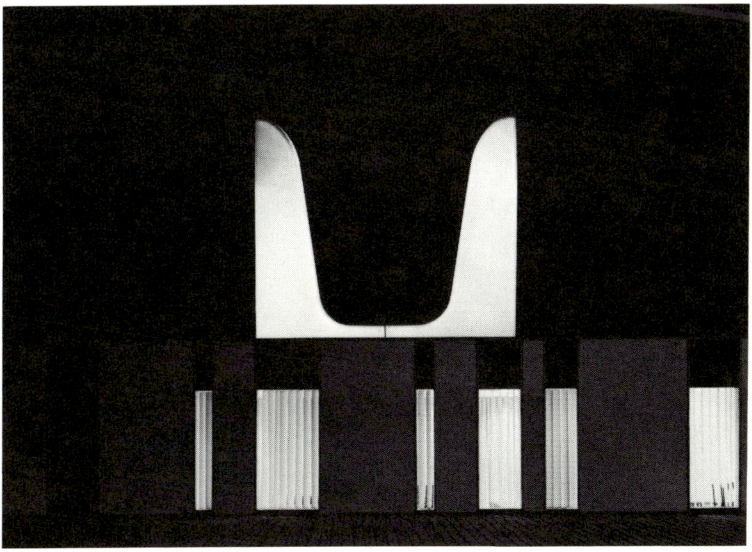

Institute of Hydrology, Wallingford, Oxfordshire. Designed by Architects Design Partnership, 1971. (*COI*)

The Royal College of Physicians, Regent's Park, London NW1. Designed by Denys Lasdun & Partners, London, 1960. (*Brecht-Einzig*)

Luxury flats at 26 St James's Place, London SW1. Designed by Denys Lasdun & Partners, London, 1958.

Store and offices for Peter Robinson Ltd, Strand, London. Designed by Denys Lasdun & Partners, London, 1958.

Leckhampton House, Corpus Christi College, Cambridge. Designed by Arup Associates, London, 1965. (*Edward Leigh* above, *Colin Westwood* below)

The Priory, Blackheath, London. Designed by Eric Lyons & Partners, Surrey, 1955–6. (*Owen Lawrence*)

Leicester University Engineering Building. Designed by Stirling & Gowan, 1959. (*Brecht-Einzig*)

British road sign by Jock Kinneir, 1962, who states 'that the lettering is only a part of the job, and that the layout, based on a series of minimum preferred dimensions, is every bit as important'. Green background, with white lettering for towns and yellow for road numbers. (*Jock Kinneir*)

XV

The British European mind questions frequently and compares compulsively. It finds both the best and the worst in Britain unequalled anywhere in democratic, industrialised countries and is acutely conscious of the disproportion in our country between the average, which is bad and very much in evidence and the good which is rarely seen because there is so little of it.

British Europeans see themselves as Britons first and Europeans afterwards; they are either British or Europeans by choice and this implies the recognition of differences; it involves judgment and ultimately action. They can therefore, by definition, not stand aside. They actively advocate those changes which will, to their minds, benefit a European, Western-orientated Britain. This avoidance of the neutral, non-committed stance is possibly not very British but it may nevertheless be quite important to Britain.

I have looked, in this book, at today's Britain with British European eyes through aesthetically coloured spectacles set in an industrial frame. It seems to me that it might be good if we were to spend slightly more time in looking forward rather than backward and slightly more effort in learning from European and other Westernised nations rather than in trying to persuade the unwilling, backward nations of the world to learn from us.

I am also fairly clear in my mind that many of our social and economic problems would be overcome in, say, fifteen or twenty years' time if certain changes were introduced in our educational system now so that today's children could evolve a better sense of visual and absolute values which would enable them to assess and to relate what they have to give with what they want to get.

In the last analysis, design is concerned with people and with their sense of dignity and the acid test of each aspect and of the sum total of design is whether they enhance human dignity or detract from it.

Good design uplifts the soul and helps people to feel better –

about themselves, about others and about the world; it makes it easier for people to make the best of themselves, helps them to realise their potentialities and so raises their personal stature. Indifferent or bad design usually has the opposite effect; it makes people feel depressed, foolish, out of place and wrong and so diminishes the individual.

Human dignity has not, so far, occupied a very high position in our scale of conscious values. Yet none of the enormities of the past fifty years could or would have been committed had each new departure into uncharted philosophical and political and economic areas and their practical implications for the individual, been put to this simple test: will this attitude and its immediate practical effects enhance individual human dignity or detract from it? Good design defends dignity just as surely as bad design destroys it.

To design means to care about people as individuals. Or, perhaps, it means simply to care.

APPENDIX A1

The total number of students in art establishments, showing the proportion of full-time, evening and other part-time 1959–1970

From *After Coldstream*, Piper (Ed.)

The total number of students in art establishments, showing the proportion of full-time, evening and other part-time 1959-1970

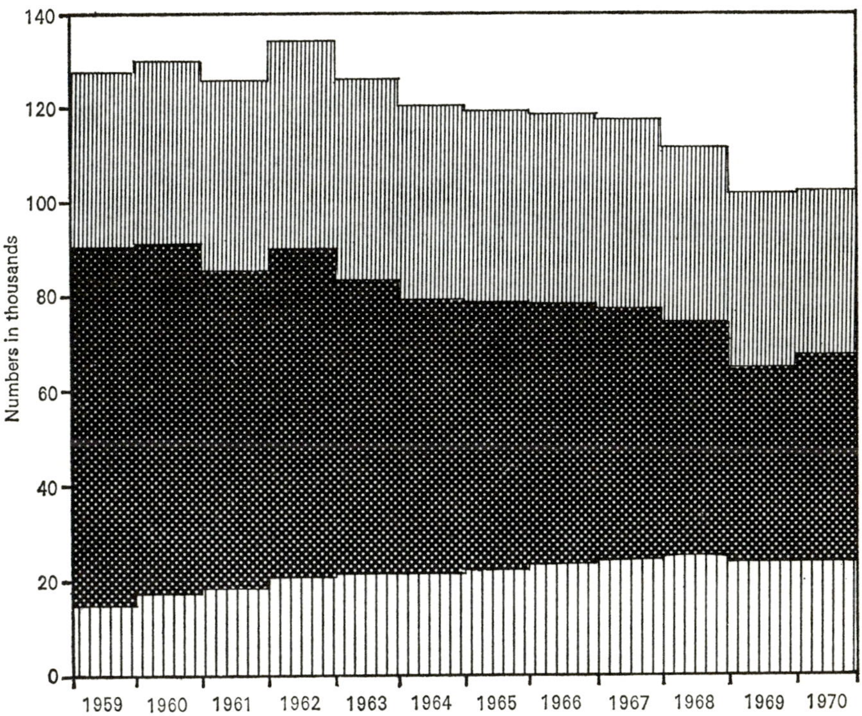

Key

Other part-time

Evening

Full-time

Source:

Education in
1959 Table 59
1960 Table 55
1961 Appendix B

Statistics of Education
1962 Vol 2 Table 9
1963 Part 2 Table 5
1964 Part 2 Table 5
1965 Part 2 Table 15
1966 Vol 3 Table 1
1967 Vol 3 Table 2
1968 Vol 3 Table 2
1969 Vol 3 Table 2
1970 Vol 3 Table 1

APPENDIX A2

Summary of the distribution of art and design students in Further Education in 1970

From *After Coldstream*, Piper (Ed.)

	Non-advanced courses leading to recognized qualifications				
	OND ONC	City and Guilds	GCE	Other non-advanced courses	**Total**
Art Establishments					
Full-time	6	1,552	580	10,100	**12,238**
Sandwich	—	83	—	—	**83**
Short full-time	—	—	—	—	—
Part-time day	124	10,415	1,137	501	**12,177**
Day release	—	—	—	—	—
Evenings only	18	1,609	3,458	714	**5,799**
Art Estabs. **Total**	**148**	**13,659**	**5,175**	**11,315**	**30,294**
Polytechnics					
Full-time	—	—	—	—	**1,492**
Sandwich	—	—	—	—	**76**
Part-time day	—	—	—	—	**107**
Evenings only	—	—	—	—	**22**
Polytechnics **Total**	—	—	—	—	**1,697**
All grant aided establishments **Total**	—	—	—	—	**31,991**
As a % of total for all courses	—	—	—	—	***31·2***

[1]Includes the London College of Printing, not included in the column above

Advanced courses leading to recognized qualifications							Total all recognized courses	Total un-recognized courses	**Total all courses**
Dip.AD	ATD ATC	Other national qualifications	Professional qualifications	College Diplomas and certificates	Other courses	**Total**			
3,959	280	963	447	350	83	**6,082**	18,320	743	**19,063**
—	—	55	—	73	—	**128**	211	—	**211**
—	—	—	—	—	—	—	—	24	**24**
—	—	138	58	—	43	**239**	12,416	10,433	**22,849**
—	—	—	—	—	—	—	—	11,253	**11,253**
—	—	98	81	—	81	**260**	6,059	37,488	**43,547**
3,959	**280**	**1,254**	**586**	**423**	**207**	**6,709**	**37,006**	**59,941**	**96,947**
3,174	320	—	—	43	64	**3,601**	5,093	—	**5,093**
—	—	—	—	13	—	**13**	89	No	**89**
—	—	—	—	—	—	—	107	published	**107**
—	—	—	—	—	10	**10**	32	figures	**32**
3,174	**320**	—	—	**56**	**74**	**3,624**	**5,321**	—	**5,321**
7,320	**600**	**1,254**	**586**	**479**	**281**	**10,520**[1]	**42,327**	**59,941**	**102,268**
7·2	—	—	—	—	—	***10·2***	*41·4*	*58·6*	***100·0***

Source:
Statistics of Education 1970 Vol 3 Tables 1, 7, 10

APPENDIX A3

Some Figures about Art and Design Education From *After Coldstream*, Piper (Ed.)

	1960 Number (thousands)	% of total	**1970** Number (thousands)	% of total	**Growth** % of 1960 number
Evening classes	73·7	57	43·6	43	– 40·8
Part-time day/ Day release	37·6	29	34·2	33	– 9·0
Full-time	17·8	14	24·2	24	+ 36·0
Total	129·1	100	102·3	100	– 20·8

APPENDIX B

Selected Extracts from *After Coldstream*, Piper (Ed.), Book 2 of Readings in Art and Design Education by Francis and Piper

The number of local authority art establishments has declined annually from 180 in 1959 to 125 in 1969 and 111 in 1970. 77 of the current establishments are colleges in their own right, the rest are departments within colleges. Most art teachers' training courses are also run by local authorities, sometimes attached to an art college; the total number of art training centres was 16 up to 1966 and 13 since then.

Student numbers

In 1970 there were a little over 100,000 students attending art establishments in England. Evening class students made up 40% of the total, part-time day and day release students made up 33% and full-time students 27%. In 1960 57% of the students were in evening classes. 29% on part-time day courses and only 14% were full-time. This figure shows an overall decline of 20% but the decline is really only in the evening class and part-time day students whereas full-time students have gone up by 36%.

Trends in numbers taking various courses

Only 10% of full-time art students were on courses leading to advanced national qualifications.

Art Teachers Diploma and Certificate

The qualification for art teachers (other than the BEd given by universities) is the Art Teachers Diploma or Art Teachers Certificate. Although three art training centres have been closed down in the last ten years the number of students on these courses has not dropped.

Foundation Courses

Pre-diploma courses are now renamed 'Foundation Courses' to indicate that students completing them are not assured of a place in a diploma course. In 1970 there were two Foundation Course students for every DipAD place given. There are no figures to show how successful applicants from Foundation Courses are compared with other courses or who have no Foundation Courses at all; but assuming that the acceptance rates of the two groups are similar, about 42% of the 1970 Foundation students were on to DipAD in 1971.

Conclusions

Statistics show three major developments:

1. Rapid contraction in the number of art establishments with an attendant concentration on courses leading to advanced qualifications at the expense of non-advanced courses.
2. A fall-off in the number of evening students which does not reflect the general trend.
3. A growth in the number of full-time students which has kept pace with the population growth at a time when further education as a whole has been accommodating a growth proportion of school leavers in full-time courses.

Over the ten years there has been an attempt to improve the quality of advanced art and design education by the introduction of the DipAD and the establishment of the National Council for Diplomas on Art and Design. The restriction of DipAD to those colleges judged by the NCDAD to be of sufficiently high calibre during its first years of operation has ensured that any improvement of quality has been achieved at the expense of quantity.

It seems reasonable to assume that art education is falling behind the demand that there is for it; because art education has expanded at a much lower rate than further education has done as a whole, and there is no reason to believe that the demand for art education has not increased in the same proportion as the demand for education in general has.

The National Structure

The importance of the educational system as a determinant to our future society is greater now than it has ever been. Since all art education is state-run the critical decisions are in proportionately fewer hands, but education is costly: it is one of the government's largest bills, £3,000 million in all. Therefore we cannot afford to make mistakes.

There are three possible ways of looking at the question of spending money on education and of saving money on it:

1. To limit educational growth to keep it in step with the growth of the Gross National Product.
2. To meet a small proportion of the education costs from public funds, leaving more to private sources.
3. To make the educational system more efficient in turning public money into public benefits.

The first suggestion is difficult to carry out because it would mean abandoning the principle which was elucidated in the Robbins Report and which was for providing courses for all those who are qualified by ability and attainment to pursue them and who wish to do so. The second choice is unrealistic for obvious reasons. It seems clear, therefore, that we have to try and make the educational system more efficient, in other words, to adopt the third course, and here there is plenty of room for improvement.

There seem to be three major areas where one could make our system more efficient. The first is to cut back on overhead expenses, rationalise the use of buildings and equipment and organise educational institutions so as to get the best economies of scale. The second would be to seek ways of increasing the student-staff ratio without dropping standards. This can only be done by improving the quality of the staff and by adopting teaching procedures which allow larger numbers of students to benefit from each member of the staff. Part-time staff represents one kind of a problem and full-time staff presents quite a different kind of a problem and because art colleges rely on both, both kinds of staffing have to be very carefully considered in this connection. The third area where change would help to make the system more cost-effective is its responsiveness to change:

'The eventual aim should be to produce an educational system which can be entered in a variety of ways by students with different levels of attainment. Once inside they should be presented with a comprehensive choice of courses, enabling them to pursue their interests, be it training for a job or self-development (and they are not mutually exclusive), to the full extent of their ability, or as far along that increasingly difficult path as their own motivation will carry them. The system should seek to create opportunities for people, not categorise them as soon as possible and then present them with a package deal on a take-it-and-pass or leave-it-and-fail basis. It should be possible for a student to reach any of the qualifications on offer from whichever point he enters the system, not as an exception (as when a person transfers from a vocational to a DipAD course), but simply as a person who has entered at a level appropriate to his previous experience and taken a normal path through the system.

'The concept of a network of courses has grown as a response to this kind of educational thinking. A network system should not be confused with a cafeteria-style system in which students help themselves to courses catching their fancy. Nor is it synonymous with "generalist" education. A network simply relates courses in such a way that together they form a greater number of sensible paths along which a student may progress than would be the case if they were run independently. Instead of starting from an idea of what a course should be and then selecting and rejecting individual students in accordance to how they fit, it seeks to provide the right mix of courses to fit the mix of people coming forward; it has a positive alternative to offer a person who fails on a course. Further, it gives greater choice to those who successfully complete a course. It may be arranged so that decisions which exclude a student from a particular course are taken as late as is possible without jeopardising necessary standards. Most important of all, such a system is adaptable: courses which attract large numbers can grow, those which do not work well can be modified or wound up. Resources can be shifted where they are needed. The whole can adapt to changes in student intake, or economic need. It is, above all, a dynamic system.'

Who decides what is to be taught?

At the moment there seems to be nobody who is responsible to the public when it comes to the forward planning of what is being taught at art colleges and polytechnics. The UGC (University Grants Committee) had a planning function and in earlier days much of its discussion was concerned with policy. Today the UGC has no right to consider polytechnics; there is no public committee for the polytechnics which corresponds to the UGC. All planning seems to be done by the Department of Education and Science, and this department opposed the setting up of its own Planning Branch, it resisted the creation of the Council for Education Technology and it does not like the Central Advisory Council. This means that the DES is keeping policy-making behind closed doors.

Art Education has two committees: the National Advisory Committee on Art Education which is no more than a collection of people who can, when they are asked for it, give an informed opinion but little else; and the National Council for Diplomas in Art and Design, now incorporated in the Council for National Academic Awards. This body is concerned only with the academic standards of schools and colleges awarding degrees, and cannot concern itself with any other matters.

Far from there being a body which can take an overall view of education or of further education there is not even a body which can concern itself with an overall view of art and design education.

It is argued that this is the job of the DES; in that case, its deliberations should be made accessible – which they are not.

Neither the politician nor the civil servant is in a position to tackle fundamental problems like the function of the education system or the relative merits of possible national administrative systems. There seems to be no way acceptable to civil servants in which they can be involved or even informed by discussions with people from the outside. This does not help create the correct policy of education for the country. The DES does have a Planning Branch which from time to time produces study papers, but who suggests what the DES Planning Branch should do? Whoever it is, is he of the right calibre, experience and training to tackle fundamental questions of policy? Even members of the NACAE, advising the government on education, are left in the dark over current DES thinking.

APPENDIX C

Secondary School Education

To gain an idea of the rôle which art and crafts plays in our secondary schools and of the mentality of those who determine future trends in these subjects, it may be worthwhile to look at the *Report of the Working Party on Art in Secondary Schools* prepared for the Scottish Education Department.

The Working Party was set up in 1968 and its terms of reference were:

> To consider the position of art in the secondary school in providing a continuing experience in creative expression through the visual arts and in fostering a sense of responsibility towards the creation of a visually satisfying environment.

The Working Party reported in 1971 and considered, first, the position of Art in the Educational system and in society during the past two decades.

The introduction of the Report sets the scene, looks at the present and suggests some developments for the future. The Report itself makes observations and recommendations on specific aspects and levels of secondary art education; it looks at the first two years of the secondary-school course and on their relationship with the last years of primary-school training. In its review of the years which follow the first two years at secondary school it deals with the three broad areas of involvement in the visual arts.

In the introduction itself certain views are expressed:

> Throughout the course of history painters and sculptors and poets have often been among the first to recognise or initiate and give expression to far reaching changes in spiritual, intellectual and formal values. It is not surprising, therefore, that in the most recent history of education, art teaching at its best has probably made one of the major contributions to the development of a system which is concerned for the development and well-being of the individual.

Art is a subject in which the very old can be related to the very new when in fact the understanding of what has been achieved may be the key to what is being now attempted. It is an area of study in which, for the great majority, understanding is enhanced by personal participation, however modest the level. Art is concerned largely with individuals and with individual judgments.

As artists, children are different from but in no way inferior to adults. Artists themselves were among the first to recognise the rich fantasy and vivid powers of expression in children's art and to direct attention to its educational significance.

The economic and social climate of Scotland has probably contributed much to public reticence regarding the visual arts in society and perhaps partially explains the curious situation in which literature has enjoyed much greater acceptance in our cultural and educational life. A genuine appreciation of children's artistic work is still not widespread. All too often, parental interest and support which may be relied upon in other subjects is sadly lacking. It is not that a place for art education in the secondary curriculum has ever been seriously challenged so much as the fact that it has never been fully understood and exploited in the educational policy of our schools.

However much the level of 'public taste' may still be deplored in recent years there have been signs of greater discrimination, as reflected in many of the products available in our shops today. This has been brought about by the influence of mass media, increased opportunities for travel, the efforts of teachers, agencies such as the Council of Industrial Design and companies with more enlightened design policies.

The high degree of technical perfection in the media and the truly remarkable quality of detail and finish found in many mass-produced articles, e.g. toys, pose their particular problems to the art educationist. Today's far richer, sophisticated commercial background competes almost ruthlessly for children's interests, standards and values and makes it increasingly difficult for those who are concerned with encouraging children to tackle problems for themselves. Fortunately, however, being able to achieve results through one's own efforts is always likely to remain a strong incentive. For the adolescent, particularly, these results often must measure up to exacting standards. A situation must now be faced in which departments concerned with creative work of all kinds will require the means of dealing with sophisticated techniques.

In setting out to establish the principles and to indicate the approaches which might serve as a guide to teachers the Report continues:

The production of future executants is not the main function of the course, although the work will provide a suitable foundation for them. Increased awareness and response to the visual elements of the environment, imaginative thinking and a deeper involvement in art and crafts are within the capabilities of the majority of pupils and are the prime considerations in planning.

During the course and, if possible, as a result of an introduction made in primary school, all pupils should be encouraged to develop interests in the various art forms which they are likely to meet in adult life, e.g. painting, sculpture, architecture, photography, poetry, prose and drama.

The Report recommends that in the first two years the course in art and crafts should give a wide experience within a variety of media, that pupils should be enabled to link art activities to real-life situations and that a whole range of approaches should be acknowledged by the teachers if all pupils are to be considered creative.

It further recommends a flexible frame of mind in the art teachers during the first two years of secondary education and goes on to enumerate the elements of the course as:

1. The development of the pupil's power of observation (but not necessarily by conventional representation).
2. The importance of drawing leading naturally to graphic and three-dimensional representation.
3. Since it is impossible to determine boundaries between design, painting and craft and any other aspect of an art course in a secondary school, the stimulation and encouragement of resourcefulness, sensibility and inventiveness are prime objectives.
4. The provision of intellectual and emotional stimulation at the pupils' level of ability.
5. The help which will make pupils understand that art and design do not exist in a vacuum, but are significant factors in personal experience and in the environment.
6. Since the distinction between art and crafts is disappearing, the Report recommends planned courses of a minimum of two hours per week for the first two years of secondary education to further the development of crafts teaching and to encourage a closer relationship between materials and their sensitive use.

The Report recommends that a Foundation Course in art, begun in the first year, should continue into the later years and should be offered to all pupils and not only to those who take a certificate course in art. As a reason for this it quotes Curriculum Paper No. 2 as follows:

In the rapidly changing society in which they live, it is their general attitude to living which will be of first importance. The moral code to which they subscribe and their relationship with other people may have far-reaching effects both for themselves and for the society which they will help to fashion. Their use of the leisure time which will certainly be theirs may have a greater influence on their well-being than the possession of an additional (often unnecessary) examination pass.

If visual art education is to fulfil its rôle in providing a highly individual extension of human experience, intellectually and emotionally, there is need for a developing course with which pupils will have contact, not only throughout S1 and S2 but in the more mature years of their secondary education.

The Report mentions some schools where one period a week is given to a general art appreciation up to sixth form level. This is not the case in most schools but even where it is the Report feels that:

The potentialities here, however, extend far beyond involvement in interesting activities in art and craftwork and offer scope for fuller understanding of the visual arts and design in society at individual and communal levels.

It recommends three types of courses:

1. Courses which provide experience, on an individual or group basis, in practical aspects of the work such as painting, sculpture, craft subjects.
2. Courses of study dealing with art, design, architecture and planning in relation to the environment, and in this area there is much common ground with other aspects of the curriculum.
3. Courses which provide opportunities to combine practical work with that which involves wider aspects of art in relation to society.

It also suggests that the teacher must try and bring pupils into contact with the visual environment and assist in developing a critical and knowledgeable approach to it and in discussion must be prepared and able to substantiate his opinions.

It further suggests that:

Relationships should be established, wherever possible, with practising artists, designers and craftsmen; with industries where common ground is apparent and where management is prepared to explain and discuss design and production considerations; with architects, planners, administrators and consumers. Visits to schools by such professional, administrative and industrial personnel and visits by pupils to outside establishments can do much to foster the interest in society and the environment which we consider a vital part of education.

There is an increasing need for public participation in the development and appreciation of the environment and it is a matter of growing importance to encourage an informed public which has concern for the quality of its surroundings. Educationists are aware of the need to cultivate an ability to evaluate not only the functional and economic but also the social and aesthetic problems which affect the environment. This is an important experience for all pupils and one which has already been recognised in primary education. (Ref. *Primary Education In Scotland* – HMSO 1965.)

Environmental appreciation should be continued and developed from primary school into the course in S1 and S2 and beyond into the later stages of secondary education. This area of the curriculum is not a matter of purely visual appreciation although it is likely that the art department will have a particularly valuable contribution to make where visual aspects are concerned. It is a process aimed at opening pupils' eyes to their surroundings and encouraging responsiveness to the elements, natural and man-made, which influence them. This outdoor work requires consolidation in the classroom as individual and group efforts and it is here that inter-departmental study will probably be most rewarding. There will inevitably be strong linkage between the arts, design, architecture, town-planning, modern studies, mathematics and science.

Throughout all the courses undertaken in this area of secondary education, the importance of an outward-looking philosophy and a relevance to twentieth-century life cannot be over-stated. As society moves rapidly into an era of greater leisure afforded by the same scientific and technological advances which are altering our patterns of living and our concepts of time and distance, art education can do much to stimulate satisfaction in personal achievement, to preserve individuality in an age which prompts uniformity and to encourage critical awareness of and adaptibility to a swiftly changing environment. Visual, artistic sensibility is too valuable a personal asset to be left to chance acquisition.

The report remarks that many pupils who might take art at certificate level are prevented from doing so because of the nature of the course options which are offered. Even without taking into account those who enter art schools, colleges of education and universities with Higher Art as one of their qualifications, for the majority of pupils the study of art could become a valuable part of their personal development whether or not it is used specifically in their future vocations.

It feels that the nature of design, its implications to society and the educational value of considering design problems are too important to be treated lightly. Design being essentially a means of achieving a purpose will frequently involve purely visual considerations in the art and craft course. Art, design and craft education must keep abreast of developments – bearing in mind that quality of work is not necessarily in proportion to the sophistication of the equipment used – and that continuous experimentation would be linked to outside contacts with industry and craftsmen.

The Report states:

> We have been thinking of the pupil more as a future, thinking consumer than as a designer but, we would hope, a consumer who by the nature of his experiences has been able to establish worthwhile criteria upon which to base his judgments. We would further hope that his awareness of his environment would be developed and some of its problems, its potentialities and the influences upon it more readily recognised and perhaps understood.

In looking at future possibilities, the Report concludes:

> Currently, a number of the traditional lines of demarcation between subjects are disappearing and this is certainly true of art and technical subjects and art and dress and design. The situation is now one which merits careful reappraisal. Wood and metal are being used increasingly in the creation of works of art and the experimental use of these and other materials has been established in art departments for some time. We are of the opinion that with the development of mechanised production techniques, the educational value of training young people in craft skills has diminished and that these skills are of less vocational significance today. It is now important that the working of wood and metal be allied to activities which provide outlets for imagination, ingenuity and personal expression. We welcome the initiative taken by some education authorities to ensure that the accommodation and equipment

provided for art departments has permitted the use of machine tools and techniques such as brazing and welding in art and design activities.

It would be false to suggest that expertise in three-dimensional design is commonly available in all art departments, but young, lively designers are to be found in growing numbers in our secondary schools today. With their background of training, they are in a position to make an invaluable contribution to this aspect of art and design education.

In relation to home economics there are, quite clearly, areas of mutual interest in textile design. Fabric printing, embroidery, weaving and dress design are of significance today to the more traditional craft skills of dressmaking, sewing and knitting. It is essential that any work in fabric crafts involving design considerations be carried out to the highest possible standards, in which knowledge of the processes involved is indivisible from a background of artistic sensibility. We have no doubt whatsoever that this can be achieved only by teachers with the depth and degree of training of the artist and the designer.

Because of the nature of the forms of its expression, almost all aspects of the visual arts require practical application or ability. Unfortunately 'practical subjects' are often associated largely with the education of the less able pupil and, as we have already stated in this report, art is denied to many of the intellectually able in the later years of secondary school.

We would be very much happier about many of the schemes for integrated study if some time were devoted to cultivating among the most able pupils a greater understanding of the important rôle which intuitive feeling and expression can play.

Some justification for the position we take may also be found in the current attitude of the university entrance boards towards secondary art education. The greater importance which they appear to attach to the history of art suggests a misunderstanding of the aims of art education.

There is no reason to question the recommendations of this Report, nor many of the assumptions on which its conclusions are based. Indeed my first reaction was one of total acceptance. On carefully re-reading, however, there is one general remark that one feels impelled to make. It is the sense of euphoria and lack of the necessary objectivity in regard to art in general and as a school subject which seems to show through the texture of the introduction.

Before substantiating these two points it might be as well to take a look at what was actually going on in our secondary schools at the end of 1973.

Of the children in Britain at that time, 99·8 percent received no formal instruction in art and crafts after the age of sixteen. More than half got none after the age of thirteen. What happens in our secondary schools in this respect matters a great deal because that is all the art instruction our children ever get. It is hoped that the interviews with teachers (see pp. 127–33) will give the reader some of the facts about as well as the mood of our secondary schools.

One's instinctive immediate reaction to this report is: 'Oh good; that's all right then.' Second thoughts, however, raise a number of questions and doubts.

In one of the concluding introductory paragraphs it is said:

> If this analysis of the situation is correct, it perhaps helps to explain why at a professional level and consequently in education, the artist is exploring with almost frenzied energy a very wide range of human imagination and experience.

It is a pity that in contrast to the Report itself, this part of the introduction exaggerates the importance of the artist and his work in society and in his relationship to it. One cannot honestly say that one has noticed artists exploring with 'almost frenzied energy' or with calm energy for that matter, a very wide range of human imagination and experience. Even less has one been aware of this kind of work being done by artists in regard to education.

Is it true that in the course of history, painters, sculptors and designers have often been amongst the first to recognise or initiate and give expression to changes in spiritual, intellectual and formal values? Would it not be more accurate to say that almost any other group of people in other disciplines, writers, journalists, architects, musicians, scientists and even soldiers and philosophers and sociologists have initiated and recognised changes more often and more quickly than have painters and sculptors? Can one think of many instances where 'art teaching at its best' has made one of the major contributions to the development of a system which is concerned for the development and well-being of the individual?

'Art,' it is said, 'is a subject where the very old can be related to the very new' – can it really? How does one relate a Cimabue to a Rothko? Or cave painting to Impressionists?

Art is said to be an area of study where understanding is enhanced by personal participation; but is not understanding in almost every sphere, and not just in art, enhanced by personal participation?

Children are said to be different but in no way inferior as artists to adults. One could well imagine serious doubts about the proposition that children are artists at all and that they can produce art in the true sense of the word. What has been recognised in recent years is that children have freshness and innocence and that their expressions, whether they are verbal or graphic, contain these qualities in larger quantities than the expressions of adults normally do; but this in itself does not make children into artists. It would, on the other hand, be quite feasible to defend the proposition that the artist combines the innocence and freshness of the way a child sees things with talent, skill and ability to conceptualise and to abstract, and with self-discipline; all of which are the results of training and experience which a child cannot possibly possess.

Reference is made in the Report to the greater discrimination of which there have recently been signs and which is ascribed to the influence of mass media, travel, better teachers, the Council of Industrial Design and enlightened companies. One wonders into which category the Hippy look, and the Carnaby Street look along with the Biba look, belong. Are these to be seriously considered as an advance in public taste? Or are they, on the contrary, the easy, lazy way out since they refuse to take up the challenges of today and retreat instead into irrelevant nostalgia for an earlier period of which they know even less than they do of their own day and age, or commit themselves to being twenty-four hours a day actors in a tragicomedy of mores and manners? Is municipal or private housing in Britain an advance on what was done in Edinburgh or in Bath one hundred and fifty years ago? Or is the blanket approval of everything that is contemporary which the Report shows, simply a tendency to confuse the facile and the trendy with the well-designed, with the carefully thought out?

It is said that because of the high finish of commercial products in general today and because comparisons will be made between the schools' efforts and these products, it is now important that the departments concerned with creative work should be equipped to deal with sophisticated techniques so that the comparisons when they are made do not turn out to be too odious. One must question the rightness of the attempt to compare the schoolwork of children to the output of commerce and industry and to use this as a lever for obtaining better means and facilities in schools. This comparison could stand up to logical examination only if both industry and schools were trying to do the same thing; which they are not. Industry, at its best, takes the creative work of designers and of artists, both equally professional, and produces them in bulk; schools should never look at their pupils as artists because even those very, very few who will become artists (·001 percent) are not artists while they are at school where they are simply familiarising themselves with the various raw materials and techniques and are copying other people's work to a larger or smaller extent; and schools should not put pupils' work into mass production.

The introduction to the Report is confusing rather than helpful because it does not help to make clear whether it is the function of art education to establish, to deepen and to widen the visual awareness of the children, to familiarise pupils with their environment and with the work of the artist or whether it is its function to make pupils into part-time, creative artists. If anything it seems to imply that it is both ('as artists, children are different but in no way inferior to adults') and this seems to me to be a highly debatable point.

Since so very few school leavers go into art colleges and since only a small proportion of these become creative artists (about one in ten thousand of all school children) it is appallingly wasteful to try to teach creation in art to all children, regardless of whether they are talented or not. There are, one would have thought, many other important things to teach them in connection with art and crafts which would be much more meaningful.

When the over-statements of the introduction are put aside, there is the Report itself which stands at a considerable distance from the facts encountered in our secondary schools today, as the interviews which were conducted made plain. Thus in discussing the planning, the organisation and the content of art and crafts courses in the later years of secondary education, that is between the ages of thirteen and sixteen, it recommends three types of courses: those which will give experience on an individual or group basis in the practical aspects of work such as painting, sculpture and crafts; those dealing with design, art, architecture and planning, with relation to the environment; and those which provide opportunities to combine practical work with that which involves wider aspects of art in relation to society.

If the position in our secondary schools is considered realistically it will be seen just how impossible it is for the same teachers (of art and crafts) as they are trained and organised today, to give children anything like adequate training in all of these three directions. Unless, therefore, there is a clearly stated aim for the provision of additional, new and different teachers to initiate these courses, nothing much will happen and nothing will change. For it should also be clear that the kind of person who can help children towards experience of painting, sculpture and crafts subjects will probably not be able to give them theoretical courses on the study of these same subjects and of architecture and town planning in their relation to the environment. The drawing and painting teacher will certainly not be the right kind of a person to give them worthwhile courses which will combine practical art and crafts work in relation to society. One really needs three different kinds of teachers for the three different kinds of courses which are recommended. No mention of this is made in the Report.

Another doubt about the Report arises from its apparent inability to follow its own recommendations. Thus the Report recommends that relationships should be established in secondary schools with artists, designers and craftsmen, with industries where common ground is apparent and where management is prepared to explain and discuss design and production considerations, with architects and planners, administrators and consumers; all this with the idea of fostering an interest in society and the environment which the Committee considered to be such a vital part of education.

If one looks at the composition of the working party one will find that the main profession represented is that of art teachers. Where, are the industrialists, the practising artists, designers and the craftsmen on the Committee? Where are the architects, the town planners and the administrators and where, above all, are the intelligent consumers? If it was found impossible to make contact with them on one committee it will certainly be difficult to work with them in thousands of schools up and down the country.

One of the most telling and thoughtful points made by the Report is that throughout the courses which they recommend for secondary schools, the Committee has been considering the pupil more as a future, thinking consumer than as a designer; a consumer who, by the nature of his or her experiences, has been able to establish worthwhile criteria on which to base his or her judgment; in fact, a *discriminating* consumer.

One would have liked to have seen this idea followed up by the appropriate practical recommendation that a special, new kind of teacher be trained, whose sole purpose and function would be to acquaint himself and the pupils with current industrial production of objects in everyday use, which is a vast subject on its own and requires specialised knowledge and aptitudes and which they would then, together, intelligently scrutinise and discuss. One doubts if the intelligent, fully fledged consumers would suddenly materialise like butterflies out of the chrysalis at the age of sixteen or seventeen, upon leaving school, if during all their time at school they were rarely shown current industrial products and if a critical approach to current industrial products in everyday use was not a normal part of school routine. Some of the teachers interviewed did indeed say that they tried to do just this; but even in the best circumstances, it is clear from what they said the discussion of industrial products did not take up a large proportion of the syllabus – as it could not – taking account of all the other things the art teacher has to teach and do.

This is how, in summary form, Ann Robinson (in *After Hornsey*, edited by David Warren Piper) sees it:

> The present system is geared to learning facts in order to pass examinations. No constructive thinking is taught; nor are children taught to formulate individual judgments. Such a system in its basic unreality results in stress and tension and hatred of work. Logical thinking is taught as distinct from intuitive thinking in assessing, judging or creating. Today's school curriculum is weighted so heavily towards the academic subjects that Art as an important subject is totally excluded.
>
> Ideally the logical, rational chain of reasoning and the intuitive, immediate apprehension by the mind without reasoning – should both be acknowledged and developed.
>
> If the importance of designing a more balanced curriculum is acceptable, then the time allotted to Art should be considerably increased and the object of teaching it clearly defined so that the importance of its aims should be more generally recognised and understood.

She suggests that one way of designing a more balanced curriculum would be to allot an equal division of time to the three major groups of subjects, thus achieving an equilibrium between the various emotional experiences:

Group 1. Sciences and Maths (in these subjects the use of Logic is dominant).

Group 2. Subjects based on the study of Man and the World (Logic tends to be important because of examination pressure, though the vital part played by Intuition in creative thinking even in this field is increasingly recognised).

Group 3. The Visual Arts (where creative problems have no learnable answers and demand insight and intuitive thinking. This inevitably involves intuition – logic – emotion).

The division of time between the more logically demanding and the intuitive subjects would certainly provide more balanced mental exercise.

In conclusion, one cannot help commenting on the Report's inclination, not unique by any means, to obscure the dividing line between the copying of and so acquainting oneself with other people's art, and creating art. This attitude has led to (or perhaps stems from) a similar lack of clarity in our schools and shows up in the interviews with art teachers. It is important to try and give the term *creativity* its correct meaning for, by defining it, we might understand what children want to do and what they are capable of doing; we might then save ourselves the trouble of trying to get them to do things of which they are manifestly incapable. This, in turn, should enable us to utilise our educational resources and the children's time far more purposefully and effectively.

APPENDIX D

On specialists versus generalists at our schools of art and crafts

The NACAE (National Advisory Council on Art Education) and NCDAD (National Council for Diplomas in Art and Design) Joint Committee was a result of the student revolutions at Hornsey and Guildford in 1968–9, which happened because over the previous ten years or so the concept of Fine Art had changed: Pop Art is a very different thing from Fine Art in the old sense; craft-based industries had become automated and demanded technical knowledge which the art schools were not providing; and students felt that such instruction as was available was irrelevant to the needs of the day.

There is now a tendency to look at the newly emerging shape of art schools in the light of the following points highlighted by the art school students' demonstrations:

1. There should be no dividing line between the so-called Fine Arts and Design, and compartmentalised education is therefore irrelevant. Students now want a 'general' education. They feel that specialisation is undesirable in this era of accelerating technological development as techniques studied at school will be outdated before the student can apply them.
2. Studies should be based on a network curriculum allowing students to study those subjects which they themselves decide best suit their development.
3. The aim of Art/Design Education should be to produce generalists not specialists.
4. All general-education qualifications should be eliminated as a requirement for entry into an Art College.
5. That the students' point of view should be expressed whenever decisions are made.

When these tendencies are analysed it is found that the essence of the debate which continues today in regard to the art schools is nominally, at least, concerned with producing specialists as against generalists.

Sir Misha Black and many others who are in favour of training specialists believe that, at their extremities of maximum achievement, Art and Design are different activities; activities sharing only creativity and some techniques. Art is expressive of the human condition, it provides clues to what we cannot explain in rational terms and can achieve, in Mondrian's words, 'union of the individual with the universe'. Design, on the other hand, is a problem-solving activity concerned with invention and with formal relationships. The fact that many painters and sculptors wish to play an ecological and directly social rôle and that some designers imagine themselves to be artists poses new problems to the educators but does not invalidate the view that artists should create, that is, concern themselves with the general and that designers should solve problems, that is, concern themselves with the particular.

Students, on the other hand, seem to want a network organisation which will allow them to move from painting to design, from technical to expressive experiments as they wish. This general experience, within the boundary of visual communication, will allow them later, so they say, to decide how they can best make their own contribution to social need. The students believe that specialist skill is threatened with obsolescence. A student trained broadly in the principles of design will survive, they say, while the designer who specialises in a particular technology will not.

The advocates of specialisation, however, think that the concept of undergraduate courses in which students move freely from expressive arts of painting and sculpture to the constraining disciplines of design is valid only if it leads to specialised graduate or postgraduate studies. There seems to be general agreement that rigid, linear courses are inadequate for Art/Design education and each student should be individually guided through a course of studies appropriate to his needs and ambitions. But they disagree with the network system of education and doubt if it is possible to achieve this even in a curriculum or with an internal organisation of Art and Design which is radically changed. It may be possible if courses are extended to five years; but it is very unlikely that this will happen in the foreseeable future. If only twenty percent of graduates continue to take a postgraduate study then a free unstructured network system producing generalists will produce only dilettantes; that is artists/designers without sufficient technical proficiency to support their presumptions. Only the genius could survive so profligate an educational system. On the contrary, so Black and others feel, generalisation can best develop from specialisation and only those who have experienced the rigours of applying knowledge and experience creatively to the resolving of a precise task can appreciate the reality of creative problem solution. Only students or academic theorists never faced with the need actually to solve problems advocate generalisation. Those who have been faced with the need to make positive, creative statements know that this demands not only generalised appreciation but a background of specialised skills and experience.

The defenders of specialisation say further that they know that years of dedicated study and experiment are essential before a designer can successfully deal with relatively simple problems such as the design of a plastic cup which is anthropometrically satisfactory and can be produced economically, or of an aircraft seat which is light to maximise load and yet satisfies safety regulations and comfort, with the design of truly contemporary clothes – comfortable, practical and confidence-building. If the designer is not competent to resolve the technical as well as the social and formal aspects of the problem, he is purely a stylist concerned only with the superficial appearance of the product, the manufacture of which is beyond his comprehension, and this kind of design is dilettantism (i.e. design without technique).

The task of the colleges, according to those who believe in specialisation, should therefore be to imbue the design student with an appreciation of professionalism, with the capacity to study a problem in depth, with the ability to advance from analysis to synthesis and to be sufficiently self-critical to be able to evaluate dispassionately his solution. Only the student who has mastered the control of creativity within the constraints of materials and production processes will have the capacity to influence the industry for which he later works. To begin to dominate the relationship between creativity/material/manufacturing processes/use, in one field of design, is as much as any student can accomplish in three or four years; if more is attempted, a loss of control over the medium and over oneself is the result. Generalisation alone, without initial specialisation, is no more connected with design than hospital management is with surgery or railway administration is with the design of locomotives. General and wider studies should be undertaken from specialisation but not without it or outside it. Genius is independent of formal education but a democracy cannot survive on genius alone; it requires inventiveness and human understanding as the common attributes of its middle-mass designers and requires the application of these qualities by its citizens.

Since the majority of art and design students are not geniuses and the design education in which Black believes is more easily assimilated by those who have a modicum of general education, it stands to reason, so it is argued, that certain minimum educational levels should be required by all entrants to the new degree courses. There are, of course, exceptional students who should be able to enter art schools without the normal examinations but it is important that the principals of art/design colleges willing accept these exceptional entrants and are not prejudiced by ambitions of academic respectability. It should be remembered, however, those favouring specialisation also state that, while art/architecture/design are technical subjects, they add little or nothing to the communal need unless they are also creative. The development of creativity in young men and women is dependent on their working in an atmosphere which generates creativity; this requires a continuing sense of excitement, of experiment and of discovery. There seems to be little hope, they think, of engendering this spirit in the art/design colleges now wedged into British polytechnics; nor can this spirit emerge from the schools of painting or of architecture peripheral to universities: a new start is needed. The core could be an existing school of art/design or of architecture but the concept of a real college of environmental studies must be clear and strong to engender a spirit of creative excitement.

On the face of it and in relation to their employment in industry, we train too many designers. It is arguable that to get at the very best talents, we have to explore the layer beneath the best and that, by the nature of things, this layer is thicker than the top layer and so the wasted effort in finding the best is simply part of the sifting-out process. A similar argument could be put forward in regard to the training of artists and to the relation between the number of good painters and sculptors who emerge in any ten-year period and the number who go through our art schools in the same period. This feeling of waste is shared by many designers at the top of the profession who say that, even taking wastage into account, too many artists/designers are being educated in Britain to meet even the most optimistic view of progressively greater public interest in the fine arts and of expanding opportunities for designers in industry. There are few valid counter arguments to this point of view for it seems to be based on facts; and if the number of conventional art/design students is consequently reduced or is, at least, not encouraged to grow while the facilities develop alongside other facilities for higher education, we should have a considerable amount of room for redeploying these free facilities in two new ways: we could extend and improve our adult education facilities and we could introduce new courses for a new kind of teacher.

McLuhan, among others, has argued persuasively, and it is difficult to disagree with him, that we, in the industrialised West, suffer from an amputation of the senses: of sight, of sound, of hearing and of touch; of instinctive comprehension. This applies not only to McLuhan's own country, the USA, but, as I have been trying to argue here, it is also true, to perhaps a larger extent, of Britain; but while McLuhan's point of view is a philosophical one, mine has been concerned with the effects on our industry of the ossification of the senses. To correct it we could perhaps think of using the facilities of the reorganised art schools for the purpose of adult education on a serious and large scale, so that our dormant senses are first reawakened and then reunited with our intellects. The further we move into the future the more important it will be to keep our natural audio/visual/tactile senses and methods of communication fresh and effective.

The case, in general terms, against specialisation has been put by Cornford* who quotes Charles Reich in *The Greening of America* (a book difficult to swallow as a whole, but thought-provoking in places):

'A prime characteristic of the Consciousness II person is his disclaim of personal responsibility for what his organisation does or for what his society does. He pictures himself as a person without absolute or transcendental values – he cannot make a personal judgment; he must accept the premises of society. He cites the compartmentalisation of his work, his lack of general authority. He says that he must defer to the judgments of experts in other areas where he is not competent; if

* Christopher Cornford, *After Hornsey*, Piper (Ed.).

his company is making a dangerous product, that is a matter within the special competence of Product Design; if his country is making war, that is the responsibility of the political and military experts. It is not for him to 'take a stand', he does not know enough; it would be unprofessional.'

In a different way, this has been put, by implication and up to a point, by Stanley Milgram in his book *Obedience to Authority*. On the evolutionary level too, specialisation can cause extinction and, in broad terms, adaptability – which is not possible unless there is some kind of generalisation – is an advantage.

Cornford thinks that we need both specialists and generalists in addition to our need for the 'design scientist' who is an inventor, entrepreneur, sociologist, economist, politician and industrial organiser all in one; the kind of person, in other words, of whom, by definition, there can be only a few. But, on balance, he feels that:

> People gifted in art or design are natural-born generalists; are by endowment in close touch with all the impulses of their beings; have from the first a predisposition to co-ordinate and cross-relate; to see systems and constellations where others see static-discrete entities; to perceive and act in an ecological spirit.

Blinkered specialists can often be dangerous. Cornford says that he has known several generations of young graphic designers who, for instance, despite their undoubted competence in typography, printing processes and the like, are so impervious to the real world outside their specialisation that they cannot fulfil the simple brief of designing a poster that demands to be and can be read at a distance of twenty feet and upwards. If they use the standard typeface of the day and the currently allusive smudge of an image, they seem to be satisfied; just as, one suspects, many architects are satisfied with a building that will photograph well for the magazines; if it does that, user satisfactions can be ignored.

He further feels that the reason why specialists are often blinkered is that we have made them such and the reason for that, in turn, is that there are, on the staff of most art colleges, a number of skilled individuals with a limited vision and a protectionist policy towards their particular corner of the field. To aggrandise their field, they habitually claim that its pursuit requires far more prolonged and exclusive study than it really does. Such people assume that the teacher must purvey to his students everything he knows; but can such claims be substantiated – is that the rôle of the teacher? The real problem in the education of a designer is his formation as an originator; the promotion of his critical, social and cultural awareness. The old-fashioned teacher, one who aggrandises his own field, is the one who begrudges syllabus time spent by his students on general, complementary studies. He will say: 'We are here to train designers not philosophers.' He believes in the antithesis between designer and thinker. The other kind of teacher will work on and encourage his students' involvement in general and complementary studies. The quarrels about syllabus time seem to be ridiculously archaic since general and professional studies would seem to be so related to each other as to be inseparable.

An indirect and stimulating contradiction to those who are for generalists comes from Professor Max Beloff, the historian, in a recent article (*Encounter*, November 1973) in which he reviews Charles Bohlen's book *Witness To History*. This is how he starts his essay:

> Western foreign services are generally hostile to the idea of specialisation. They regard the techniques of handling diplomatic questions as more important than the knowledge of particular countries or regions and find it hard to devise career patterns that make full use of particular expertise or training. The American State Department was amply rewarded, however, for its initiative in the 1920s, at a time when the Soviet régime had not even been recognised by the United States, in deliberately encouraging a small group of foreign service officers to learn Russian and to fit themselves in other ways for the close study of the Soviet scene. Not only does their own record of interpretation of Soviet policy show up very well by comparison with those of their political chiefs and of the services of some other Western countries, but they have also contributed by their writing to the more general understanding of the Soviet phenomenon. Historians have an especial reason to be grateful. After George Kennan's two impressive volumes of memoirs, we now have those of his friend and colleague, Charles ('Chip') Bohlen.

Political history is just about as far removed from design as any subject could be. For this reason it is refreshing to hear a historian's view on the subject of specialisation.

Mr Woodrow Wyatt, the politician, broadcaster and businessman, also thinks that specialisation pays in practical terms. In his recently published book *Turn Again, Westminster* he says:

Atlee used to advise MPs who were keen to advance to specialise. It is good advice. It is better to be thought to know a lot about one subject and thus listened to with respect as an expert, than to talk shallowly about a multitude of things. It is assumed in the case of the ordinary MP that he cannot possibly have a grasp of very much. Oddly, no one thinks it strange that the Prime Minister has views on everything or that a man who has no knowledge of medicine or hospitals should be made Minister of Health. Nor is there any surprise that when a man who has assiduously specialised gets a job it is very often in a department remote from the field he specialised in.

Nevertheless the advice stands. There is nothing like mugging up a subject for being taken seriously. It may be tax reform, the Health Service, the Air Force, agriculture, Europe, housing – it doesn't matter. The specialist must join his party's unofficial committee or group covering the subject. If he sticks at it he will become Chairman or some other officer of the group. That will give him an extra claim with the Speaker when his subject is debated, or is the occasion of questions, enabling him to establish a wider reputation.

It is, of course, quite possible and I believe even likely that, sometime in the future, a dispassionate observer of what has been happening in Britain over the past ten years will conclude that the student demonstrations and strikes of the late Sixties and early Seventies and the controversy of the generalists versus the specialists is not so much, or perhaps at all, about education and the training for careers but about something quite different.

Even now we all realise, for instance, that the so-called civil rights demonstrations of the late Sixties in Northern Ireland were not at all concerned with civil rights and liberties but with promoting the aims of the IRA behind an acceptable, respectable façade which had the maximum emotive value for the greatest number of tolerant and liberal Britons. Today, when the Catholics of Northern Ireland are thought to be quite equal to and, by some, to be more equal than the Protestants and when they are not suffering from any visible political disadvantages, the so-called civil rights campaign – after irrevocably committing thousands to its violent doctrines and methods and after thoroughly furring up the minds of many with violence, intolerance and fanatical nationalism – has crept out from behind its respectable façade and boasts of murder and arson without ever again mentioning civil rights. These are and always have been the 'salami tactics' of the extreme right and of the extreme left. They are not the tactics of honest men who spell out their reasons and intentions and means of achieving their aims as well and as clearly as they can. Similarly, it should not come as a surprise if it is found that the quantity and the quality of the changes demanded by some students and teachers in the area of higher education has more to do with the extension of the ideas of the trendy left than with a concern for people and for liberal values or with education and with its legitimate aims.

On the morning of 7th February 1974, the day on which Mr Heath prorogued Parliament because two hundred and fifty thousand miners would not respect the law of the land and the will of the House, Ronald Butt, the political correspondent of *The Times*, wrote as follows:

The worst aspect of the threatening political battle is that whether or not there is an election, the essential struggle will take place outside Parliament.

A victory is apparently to be sought in the picket lines, through industrial attrition, by conflict involving the Government and by a battle that may not be purely intellectual, for the minds of ordinary people. It is not going too far, therefore, to say that the fabric of the nation is in some danger.

To say this is not to suggest that the immediate crisis is simply the product of a Red plot. But its particular danger lies in the fact that it is happening after a prolonged failure on the part of people in places of power to understand adequately the social changes of our time and to take seriously the exploitation of these changes by minorities of dedicated people who want to overthrow the fundamental values of society by undermining confidence in them and by encouraging concepts of class enmity.

One simple illustration of how this is done is the way in which it has been possible to ridicule almost anything, from ordinary individual courtesy to a respect for disciplined learning, family life or national institutions, by a pejorative reference to them as 'middle-class values'. If the traditional communists are at all significant in Britain now, it is because a self-styled 'radical' establishment has in one way or another prepared ground suitable for them to till. One example is the manipulation for political purposes of education in schools and universities.

The other is the presentation of society not as containing injustices but as so fundamentally unjust on class lines that a new order is imperative.

This week there has been news that the deputy headmaster of Tulse Hill comprehensive school in London has resigned in despair at the failure to communicate with certain disruptive elements in that school, where the work of the majority is being, he asserts, sabotaged by small groups. In part, the trouble arises from the fact that about fifty per cent of the boys at this school are coloured and the former deputy headmaster, Mr Humber, has stated that the groups within the school with whom it was impossible to communicate were basically 'the black kids'.

Now it is perhaps inevitable that there will be some special problems in any school with so large an immigrant population, but this is by no means all there is to it. First, who is encouraging these children to be so thoroughly alienated at so early an age and before they have direct experience to justify their alienation? Why is it that so many of the people involved in race relations are apparently willing to lend substance to the assumptions that black children can expect little in the way of justice of job opportunities in their coming adult life?

But even in Tulse Hill the question is not simply one of colour or even of the result of the impersonal size of the school. This school apparently shares two problems that are widespread in others. One is an atmosphere of indiscipline; the other is the activity of groups of militant teachers who are socially disruptive among both staffs and pupils. Some of these are described by colleagues as 'neo-Trots'. They come to their schools *dressed more untidily* (author's italics) than the children; they are anti-syllabus, anti-examination and convinced of their mission to condition the children politically against the values of the society in which they live, often not scrupling to use the curriculum as well as their own social behaviour to this end. Thus, for a combination of reasons, which include the political intention of some teachers and a more widespread fashion of disparaging orderly attitudes to school work, discipline is breaking down in many places.

I have quoted Ronald Butt at length in order to show clearly that there is often more behind the demands for wholesale changes than meets the eye and that these hidden motives should, where possible, be examined. Also because he makes a point of stressing the relationship between a disruptive and destructive attitude to spiritual values and a disruptive and destructive and totally negative attitude to the personal and general, physical environment.

When all has been thought and said in favour of and against specialisation and about the real reasons which exist behind the numerous façades painted in ten-foot letters screaming for change, it is difficult to avoid the conclusion that ideally our designers and teachers need both more specialisation and more generalisation; more time at college, art school and university.

APPENDIX E

Stuart Macdonald in *After Coldstream* – on Articidal Tendencies (A history of the ups and downs in the relations between art and design in Britain)

A Summary

According to Macdonald, the non-analytical, non-intellectual approach has always been uppermost in British fine art departments. Thus at the moment it has become fashionable to say in art schools that if you want to pass, paint Hard Edge. Twenty years ago some other fad may have been fashionable. There has always been mindless imitation. What is a completely new phenomenon in the history of art education is the contemporary articidal tendency, the death wish, the desire for the demise of the artefact, even of art itself.

Macdonald outlines the attitudes of the Government and of the art teachers which have led up to the present position. From the start the development of public art education in Britain has been an effort by the State to divert art schools from their overwhelming predilection for fine art and to throw them towards industrial art, design, craftwork and general art education. Usually things were allowed to slide until the situations were reached which, for economic reasons, caused public outcry and the Government had to intervene.

Thus during the period of the Schools of Design (1837–1852) the type of art work done in these institutions was laid down by the fine artists on the Council of the Central School and its headmasters almost all of whom were Royal Academicians and whose brotherly objective was to see that no rival institutions to their Academy would flourish. By 1849 the outcry in the Press against Schools of Design which produced no designers, brought about the sitting of a Parliamentary Select Committee. The eventual result of this was, after a period of diversion caused by the Great Exhibition, the reconstitution of the Schools of Design as Schools of Art under the direction of Henry Cole.

Cole said of himself that he succeeded in converting 'twenty limp Schools of Design into one hundred flourishing Schools of Art';

but as practically all the advanced students were pursuing fine art, they were not encouraged to sit design papers. This lack of any education in designing for and in the materials of crafts had been noted by the Technical Instruction Commissioners and in 1884 they had concluded that the Department of Science and Art should go so far as to award grants for specimens of applied art-workmanship in the materials themselves, as a test of the applicability of the design. But as there had been no outcry nor pressure from the public and from the manufacturers, who were prospering and joyfully buying and selling the most hideous designs, the Government did nothing to alter public art education. The art masters directed their pupils more and more towards the Antique and away from the needs of contemporary society.

Then Morris and his disciples took an interest in these matters and they strongly opposed art schools being devoted to the production of drawing masters and of fine artists. Lethaby advocated Schools of Arts and Crafts as 'real making shops' which were combined with museums of contemporary and historic arts and crafts which he intended to become local centres of civilisation. So, at the end of the nineteenth century, craft classes were established at the Battersea Polytechnic, at Camberwell School of Arts and Crafts, at Regent Street Polytechnic and at Shoreditch Technical Institute and then at Hammersmith School of Arts and Crafts. The greatest contribution of the LCC was the establishment of the Central School of Arts and Crafts in Regent Street in 1896 under the direction of Lethaby and George Frampton.

The art masters with their background of fine art, historic ornament and 'design on paper' looked on these new developments with trepidation. They had been brought up to believe that Fine Art led design and that design was merely a matter of passing this influence down to designers – an absurd theory still believed in, incredibly, by many art educationists. It was disturbing to hear that Lethaby and Crane were lecturing at Schools of Art and stating the opposite view, namely that if society and the general level of design and craftwork was good, art would be good, otherwise it was rotten at the roots; also that design should be assimilated through the use of particular materials and by studying actual objects made in those materials. The crisis point in the progress of British art education was reached in the years 1910 to 1914. It became embarrassingly obvious then that the Central School of

Arts and Crafts was producing designers of European reputation, whereas the Royal College of Art, the senior art institution of the Board of Education, was merely producing fine art and design teachers with a smattering of knowledge of the light artistic crafts.

In April 1910 a Board was appointed to look into the Royal College of Art and was highly critical of it, saying of the light crafts studies at the College that 'in so far as they receive a definite stylistic bent it could be described as "medieval"'. The Committee was also disturbed about the College's irrelevance to industrial design and they recommended that mono-technics should be set up in the centre of each major industry and that a student wishing to design for a particular industry should be given a grant to proceed to the appropriate centre. Thus Manchester was intended for cotton, Bradford for wool, Stoke-on-Trent for pottery, etc.

The art masters did not wish to get involved with the testing realities of using craft materials and were quite worried because they had no idea how to train designers for industry. Students taking the design paper of the Board of Education were taught to apply art to nearly finished articles; they had no knowledge of what industries were really about and it is not generally understood today that applied art as it was taught then was not applied at all, that it was not applied to an object or worked in a material; in those days applied art was design on paper that could have been applied to any particular object if somebody chose to do so.

So the art masters suggested that the older schools should remain for painters and book illustrators and that new arts and crafts schools should be established at the centres of industry.

The art masters had their way. Centres for industrial design were not established and the examination in art and design did not require designs to be executed in craft material. Designs were in the form of drawings and paintings; thus it was that it was suggested to the Government that the work of the Royal College of Art should consist largely of training fine art teachers.

In 1920 William Rothenstein was appointed Head of the Royal College and he filled it with a staff of talented fine artists and illustrators. At the Royal College the fine arts prospered but industrial design was neglected and industrial design students were 'second best'. Some students of those days thought that the College had become nothing but a fashionable fine art shop.

In 1932, due to a slump in demand for British goods, the Board of Trade appointed the Gorell Committee, which reported adversely on the cooperation between art schools and industry.

At last the Board of Education decided to implement the 1911 recommendation and established a system of regional art centres to replace the existing network system of Schools of Art. The intention was that the new Regional Colleges would not only produce 'a hierarchy of art instruction leading up to a regional Art College' but also, more important, they should become the local 'centres of civilisation' which would serve local industry by providing not only class instruction but also exhibitions.

The Council for Art and Industry set up in 1934 did its utmost to encourage this thinking. As a result the Board of Education and the Schools of Art were under pressure to promote the design production and exhibition of industrial artefacts and from 1935 facilities were provided by the Board for students to submit specimens of craftwork based upon examination designs.

From Gorell in 1935 to the establishment of the Ministry's National Diploma in Design in 1946 it was a period of increasing specialisation in designing and producing craftwork.

However the Coldstream Report of 1961 and the subsequent activities of the NCDAD resulted in a swing away from vocational, useful and specialised design education, indeed away from the everyday needs of society and towards a 'liberal' type of art education dominated by high art and tall talk. One of the accidental causes of this swing of the Sixties was the large proportion of colleges recognised by the NCDAD for DipAD in Fine Art compared with the number recognised for the Design areas. Naturally students applied for and expanded the recognised departments, thus more and more fine art staff were appointed, affecting the future structure of the colleges. By the session of 1969–70 there were nearly three thousand students following DipAD courses in Fine Art, more than twice as many as in Three Dimensional Design or in Textiles/Fashion and nearly twice as many as in Graphic Design.

APPENDIX F

The Design Council

In 1972 the Council of Industrial Design was renamed and became the Design Council. The Council of Industrial Design itself was established in 1944, when the Design and Industries Association (established in 1915) was renamed and reconstituted. Before 1915 Britain had one body which concerned itself with design: the Royal Society for the Encouragement of Arts, Manufacture and Commerce founded in 1754.

The purpose of the Design Council is to help British industry to improve the design of its products and to create an atmosphere of acceptability in the country for well-designed products. Good design is taken to mean aesthetic appeal and practicability in use and in manufacture. The initial intention was to cover the whole of industry from capital goods to fashion goods and everything in between, but fashions have been gradually eliminated from the Design Council's programmes and the emphasis in recent years has shifted to capital goods and more recently to engineering.

The founders of the Design Council realised that the industrial cycle contains the elements of supply and demand and the Council was set up to affect the nature of both. It was broadly divided into an Industrial Division to stimulate the supply of well-designed products from industry and an Information Division to stimulate demand for better design.

In more specific terms the Council was charged with the establishment and the promotion of Design Centres in different industries to be supported by contributions from industry and by grants from the Government, with the arranging of exhibitions, with the giving of advice to Government Departments and to the educational authorities and with helping in the training of designers and in the teaching of design appreciation to all pupils at school.

In theory the Council's services could therefore be considered to be these:

The Industrial Division

1. To organise industrial exhibitions.
2. To produce surveys of current industrial products and developments.
3. The compilation of an index of the best available British products (selected by panels of outside assessors).
4. A designer's selection service to facilitate the selection of the right designers for particular jobs in industry.

The Information Division

1. The distribution and the sale of films, posters, charts, photographs and books.
2. The publication of the Design Council's monthly journal *Design* which is illustrated and is available to industry and to private individuals.
3. The work of its Education Section which exists for the purpose of keeping in touch with teachers and education authorities, and has facilities for arranging lectures and exhibitions and providing teaching aids.
4. The work of its Retail Section which offers help in staff training in design, in display and in small exhibitions.
5. The work of its Exhibition Section which stages exhibitions of many kinds all over the country and also abroad, apart from the regular exhibitions at the Design Centre.
6. Gaining editorial publicity through its press office and PR Officers.
7. The use of its index library.

To enable it to perform these tasks the Council had a Grant-in-Aid of approximately £700,000 in 1972/73 which was to be increased to £870,000 in 1973/74. It also has an income of about £500,000 from other sources.

In order to gain an idea of the Design Council's work and effectiveness it may be useful to examine its latest annual report, the one for 1972/73, the 28th in the series, which does not differ radically in content and appearance from many of its predecessors. In particular it would be interesting to catch a glimpse of the Design Council at its own valuation and to examine this image factually.

The report deals first with the work of Advisory Committees and records that during the year the engineering committee met six times, established a list of 70 books on engineering design, set up a sub-committee to plan the field officer experiment, and approved the launch of an engineering design magazine.

The *Wales Advisory Committee* met twice and agreed that it was important to establish 'a presence' in Wales.

The report does not show whether the Steel Furniture Committee met during the year, but it states that 'products were regularly reviewed' and that it submitted a memorandum to the Council commenting adversely on the growth of inappropriate advertising media in the street scene.

The *Committee on Farm Buildings* met four times to discuss the feasibility of a catalogue and to prepare revised colour finishes. Three conferences on farm buildings in the landscape were held but no attendance figures are given.

The *Committee on the Gold, Silver and Jewellery Industries* contributed £2,000 for design promotional activities, and four displays were arranged at the Design Centre. It also arranged a seminar for designers and manufacturers (no attendance figures are given).

The *Committee on Retail* met four times and reviewed proposals to extend the Council's influence in the retail industry. It also discussed plans for a catalogue of services in 1974, the programme for courses and conferences, the extension of Design Centre Showcases and cooperation between the Distributive Industry Training Board and the Design Council.

The Council's *Advisory Committee on Colour* was reconstituted and reformed and a new Chairman was appointed; the Hotel and Catering Committee was wound up.

Under the heading of *Working Links with other Organisations*, four meetings are reported with designers in the motor car industry and with the Engineering Industry Training Board, all of which resulted in agreements that further meetings be held.

Close collaboration is reported to have taken place with the British Standards Institution and the report mentions that one meeting 'of appropriate staff of the two organisations was held during the year to review the arrangements for cooperation'.

The joint working party with the National Caravan Council agreed that a study was necessary on the exterior colour treatment of caravans, appointed Prof. Hardy to carry it out and planned a session on design at the NCC's next convention.

Two meetings of a working party are reported to have taken place with the British Toy Manufacturing Association and according to the report the Council continued to cooperate with the Consumers Association as well as with other organisations (no details are given).

The report then deals with the *Tour of Industry of Wales* carried out by Lord Snowdon and members of the Council's industrial staff. Sixteen visits took place in addition to meetings with the Welsh Weavers' Association and with the Principal and Lecturers of the University of Wales.

Under the heading *Service to Industry*, the report describes an experiment conducted to evaluate the Council's proposal to provide technical/design assistance to engineering companies. Four members of the staff visited 15 companies around London and arranged for six follow-up visits to advise on the design of a filter for sludge extraction from sewage, a new manufacturing process for specialist heater design and a new material required to give a specific high frequency response for radar techniques. The report concludes that since the industry welcomed assistance and was glad of the 'diagnostic' role the field officer was able to fulfil, the experiment should be continued. It also felt that a team of field officers could make a significant contribution to the identification and the solution of engineering problems.

Work was started on the compilation of a record of courses of design expertise (presumably in regard to engineering) and the Designer Selection Service, established to help companies to find suitable industrial designers, handled 356 enquiries (20% up). No figures are given on the ratio of appointments against enquiries.

The report then deals with the awards which the council makes annually. Twenty-one such awards were announced for industrial products, 23 awards were given to 25 posters from the 466 submitted, 428 guests attended the seminar and lunch which, along with the awards scheme, is sponsored by organisations representing design, advertising, printing and environmental interests.

Under the heading of Conferences, Courses and Seminars 32 events were organised compared to 27 in the previous year. They are summarised as follows:

Management

Conference for Display Managers held in Amsterdam in April 1972. Attendance 33.

Management of Design for European Markets. London, May 1972. Attendance 515.

The Design Process. London, October 1972. Attendance 14.

Planning for Improvement in Engineering Design. Cardiff, October 1972. Attendance 48.

Management and Engineering Design. London, November 1972. Attendance 62.

Management of Innovation. London, November 1972. Attendance 48.

Standardisation and Engineering Design. London, November 1972. Attendance 68.

Architects and Planners

The use of colour for Farm Buildings in the Landscape. London, May 1972. Attendance 45.

Study Tour of Southern Europe. May, 1972. Attendance 15.

The Specification of Floor Coverings. Wolverhampton, October 1972. Attendance 30.

Planning Pedestrianisation and Street Furniture in Europe. Leeds, November 1972. Attendance 96.

Structures in the Landscape. London, November 1972. Attendance 54.

Space Needs in the Home. London, November 1972. Attendance 21 + 37.

Colour Technology. Manchester, February 1973. Attendance 43.

Engineering and Industrial Designers

Human Values in Design. London, April 1972. Attendance 20.

Design in Industry. Newport, June 1972. Attendance 20.

Colour Courses for Carpet Designers and Colourists. London, September 1972. Attendance 27.

The Design and Use of Symbols. London, September 1972. Attendance 84.

Courses for Retail Buyers and Staff. Toys Course. London, April 1972. Attendance 17.

Furniture Course. Oxford, May 1972. Attendance 11.

Buyers Design Course. Liverpool, May 1972. Attendance 19.

Bathroom Design and Planning. Oxford, May 1972. Attendance 8.

Selling and Display. Harlow, September 1972. Attendance 19.

Selling Toys. London, September 1972. Attendance 18.

Furnishing and Dress Fabrics. Manchester, October 1972. Attendance 39.

Household Textiles. Manchester, February 1973. Attendance 26.
Furnishing and Dress Fabrics. Manchester, February 1973. Attendance 40.
Display for Retailers. Tring, March 1973. Attendance 20.

Educational Activities

The Leverhulme Travelling Scholarship in Industrial Design, Design Council Administered Award – entries from 21 colleges.

Careers in Industrial Design

Advice was given in lectures (no statistics available). The first comprehensive directory was published to enable students to search by subject or region.

Warboys Memorial Trust Fund

Applicants from four colleges of Art and Design were interviewed and the candidate was selected.

Design Technician Courses

The Council submitted a memorandum to the Department of Education and Science Technician Working Group.

Other Education Work

During the year 617 groups comprising 9,117 people visited the Design Centre and of these 261 Secondary School Groups and 94 other groups were given short introductory talks. Over 2,500 enquiries from school children, students and teachers were answered. Enquiries to the Council's Lecture Panel resulted in 175 talks being given.

The Design Centre

The year's attendance was 1,766,456 (1,539,398 in previous year). 4,954 exhibits were shown at the centre, 6,318 trade enquiries and 6,849 personal enquiries were dealt with. The souvenir shop continued to improve on its early promise.

Design Centre Display

There were 53 special displays at the Design Centre in addition to the thematic exhibitions of which there were eight and overseas exhibitions of which there were six.

Retail Displays

Six displays were installed in showcases at Harlow each for a two-month period.

Design Index

The Council's record of well designed British goods in production contained details of 9,941 items from 1,875 firms with duplicates in Glasgow and Liverpool.

Design Centre Label

Sales of the label amounted to 13,465,000 out of which 1,950,000 were the Steel Appeal Label.

Publications

Circulation revenue for *Design* magazine rose as a result of the increase in the price from 25p to 35p. But circulation went down to 17,796 from 18,196. Paid advertising increased by 5% and revenue by 24%.

An Editorial Advisory Board was set up to advise on publications on engineering design.

Four new titles, *One Room Living*, *Camping and Caravans*, *Selling Pottery–Glass* and *Selling Floor Coverings* were published, as was *Designing for European Markets*, based on papers from the Conference on The Management of Design for the European Markets.

One magazine, *Management and Engineering Design*, was produced. A directory of Design Courses in the UK was published as were revised versions of several existing general and provisional publications.

Publicity

Press coverage amounted to 4,519 mentions; 28 press conferences were organised, coverage of Council activities on radio and television continued to be good.

The Scottish Committee

Attendances at the Scottish Design Centre in Glasgow continued to grow. Fifteen major exhibitions were held and a number of minor ones.

Industrial Activities

The Engineering Advisory Group met regularly and promoted a course on Industrial Design in Light Engineering (attendance eight). On Marketing and Engineering Design (attendance 30).

An exhibition was held on Designers in Scotland and a Conference held on Buildings in the Landscape (attendance 165).

Retail Activities

The summer exhibition on Shopping in Britain was a joint promotion with four Glasgow retailers, two short courses on Living with Design were attended by 45 housewives.

Education

Sixty-one groups were given conducted tours. Scottish press, radio and television coverage continued to be exceptionally good. Circulation of the committee's Newsletter (bi-monthly) was increased.